WJEC & Eduqas

A Level
Biology

*A Student Guide
to Practical Work*

New Edition

Marianne Izen

Published in 2017 by Illuminate Publishing Limited, an imprint of Hodder Education, an Hachette UK Company, Carmelite House, 50 Victoria Embankment, London EC4Y 0DZ

Orders: Please visit www.illuminatepublishing.com
or email sales@illuminatepublishing.com

British Library Cataloguing in Publication Data

A catalogue record for this book is available from the British Library

ISBN 978-1-911208-22-8

Printed by Ashford Colour Press, UK

06.22

The publisher's policy is to use papers that are natural, renewable and recyclable products made from wood grown in sustainable forests. The logging and manufacturing processes are expected to conform to the environmental regulations of the country of origin.

Every effort has been made to contact copyright holders of material reproduced in this book. If notified, the publishers will be pleased to rectify any errors or omissions at the earliest opportunity.

This material has been endorsed by WJEC Eduqas and offers high quality support for the delivery of WJEC Eduqas qualifications. While this material has been through a WJEC Eduqas quality assurance process, all responsibility for the content remains with the publisher.

Editor: Nic Watson, Haremi

Text design: Nigel Harriss

Layout: York Publishing Solutions Pvt. Ltd.

Permissions

Picture credits: p19: DenGuy/iStock; p21 (top): Alina Cardiae Photography/ Shutterstock.com; p21 (bottom): luchunyu/Shutterstock.com; p29: TachePhoto/ Shutterstock.com; p30: Don Fawcett/Science Photo Library

Acknowledgements

The author and publisher wish to thank Dr Colin Blake and Kevin Davies for their review of the book and their expert insights and observations.

The author wishes to thank Valerie Burden and Manjiri Joshi for the photography.

Contents

Introduction

This book will guide you in preparing for the assessment of your practical work in AS and A Level Biology for WJEC and Eduqas. Your main sources of information and help are your teachers, but it is important to be self-reliant and to know what your specification requires. Use textbooks and study guides and take responsibility for your own learning. Make it a point of principle never to leave a lesson if there is anything you do not properly understand. Many students do not realise that developing the habit of asking questions is an important way to learn and to test the limits of their own understanding. So keep on asking.

 Pointer

If you haven't understood something, the chances are that others haven't either, so never be afraid or embarrassed to ask. It shows you are thinking and by asking, you are articulating biological concepts. The more practice you have at that, the better.

Why we do practical work

You can read about experiments, watch a video of a dissection or look at photographs taken down a microscope. But, if you do it yourself, you will learn much more. For example, you will grasp the importance of bringing solutions to the right temperature for a reaction, you will know the feel of adipose tissue and realise how it affects adjacent organs, and you will appreciate the relationships of tissues within a structure. You may also make original observations. This is why we do practical work in Biology and why you should take every opportunity presented to you to do these things.

Pointer

You learn some things by reading, like how to classify enchanter's nightshade. You learn some things by watching, like the courtship behaviour of the great crested grebe. You learn some things by doing them, like culturing bacteria using aseptic technique. You learn best by doing.

How this book is organised

This book is organised in sections that address the various types of practical activity that you will undertake and how your work will be assessed. This book outlines the methods of assessment used by both WJEC and Eduqas. Each section of the book gives advice about how to undertake the tasks and how to present them when you prepare your work for assessment. You will find useful pointers in the margins. There are also suggestions for resources to support your practical activities and, because spelling, punctuation and grammar matter, a section about appropriate use of language in describing your work.

Outline of what is assessed

You will be assessed on your microscopy skills and on how you design an experiment, how you analyse and evaluate it. This book explains in detail all the areas you have to consider and gives advice for each. It explains clearly how to present all this in a report. The criteria by which practical work is assessed are described on pages 81–83. There are two parts to the assessment:

- In both AS and A Level Biology, written examinations will contain questions about practical work. The types of question you might be asked are described on pages 79–80.

- WJEC assesses practical work in a practical examination near the end of the course and Eduqas awards a practical endorsement. The skills required for these are explained in Section D: Assessment of practical work.

A note on health and safety

The experiments that you do may use corrosive chemicals, such as acids. You are likely to use enzymes, which, being proteins, may elicit a skin reaction. Alkalis and hydrogen peroxide, common reagents, are highly damaging to eyes. Methylene blue will turn you blue, and iodine solution will turn you brown. The list can go on, but it is enough for you to realise that great care must be taken with lab chemicals. Keeping the skin covered is useful but if you do get chemicals on your skin, wash it immediately under running water. Goggles are an essential precaution in some experiments.

You may need to heat a solution in your experiment. If you use a water bath, there is the worrying proximity of water and electricity. If you use a Bunsen burner, there is a flame to worry about and if you are also using ethanol, there is a risk of fire without the usual precautions. Take care at all times and follow standard lab procedure.

When you are given animal material for dissection, it is likely to be a mammalian organ or a whole animal. Some people like to wear disposable gloves, but whether you do or not, wash your hands thoroughly afterwards. While you are dissecting, always point the scalpel blade away from you. Never have your friend hold the material still while you cut. Other people's hands must be well away. If by any chance you do cut yourself, or your friend, hold the skin under running water for five minutes and consult your school nurse.

Always think what you are doing. Follow lab rules and stay alert and aware.

● ● ● **Practical tip** ● ● ●

Even your microscope is fraught with danger. Trailing electrical leads can be dangerous. If you pick up the microscope in the wrong way, such as by the eyepiece tube, you risk dropping it. Remember that school microscopes may well cost several hundred pounds each. Don't break them.

A: Practical techniques

A.1 Microscopy

When Year 7 pupils use a microscope, they forget to unwind the power cord completely so there is a shadow across the field of view. Then they put the slide on the stage at a strange angle. Then they touch the eyepiece lens and leave their fingerprints behind. Then they say that the microscope doesn't work. It doesn't have to be this way.

Setting up

Correct placement of microscope and body. Your chair must be the right height so that you are not stretching or leaning over. If you are right handed, place your page to the right of the microscope and use your left eye. Keep your right eye open so that when you draw, you can superimpose the images and draw around what your left eye is seeing. It takes some practice.

Unwind the power cord completely and make sure it is well away from the light source.

On modern microscopes, the tube holding the eyepiece lens is angled. It points towards you when you use the microscope. The power cord comes out at the back so it does not get in your way. Put the microscope a few cm from the edge of the bench, symmetrically, and if you need to, stand up to look down it, rather than grabbing the tube and pulling it towards you.

Use the coarse focus control to move the stage to its lowest position.

Put the slide on a blank piece of paper so that you have an idea of the actual shape and size of the specimen. Then put the slide on the stage with the specimen placed so that you can see the light from the condenser passing through it.

● ● ● Practical tip ● ● ●

Clean the eyepiece and condenser lenses with lens tissue before you look down your microscope and then make sure you do not touch them.

Objective lens

Start with the ×4 objective lens. Raise the stage gradually until the image comes into focus. Two possible problems:

1. If the specimen has very pale stain, like some preparations of red blood cells, you may pass straight through the plane of focus and miss it. So move the stage slowly and don't blink.
2. If the specimen is hollow, like some plant stems, the solid part of the specimen may be outside the field of view. If this happens, you won't see anything. So move the slide slightly in all directions until you find the specimen.

When you have the specimen in focus on ×4, what you do next depends on the specimen, what you want to look at and the detail you want to see. You can increase the magnification of the image by using a ×10 and then a ×40 objective and, if you have oil-immersion lenses, ×90 or ×100 objectives. Always go up in order. If you cannot focus on an image, go back to the previous objective lens and find it there first. Then rotate the nose-piece which holds the objectives, but without touching the focus control. The image should be almost exactly in focus as you go up through the magnifications.

When you use a ×40 or higher magnification objective, remember to use the fine focus control, not the coarse control. If you use the coarse focus control, the stage will move too far and you are likely to break the slide and scratch the objective lens. Best not to.

Light

Most school microscopes do not allow you to centre the condenser, but when the microscopes are serviced, this should be done. But if you can centre the condenser, do so because it will improve the image.

When you have the object in focus, to improve the image quality, you can adjust the light intensity. Some microscopes have a disc below the stage which has circles of different sizes that let through different amounts of light. You can rotate the disc to select the circle that gives you the best image.

Most microscopes have a sub-stage iris diaphragm. You can adjust this to give the light intensity which produces the image you find most comfortable to look at. Ideally, the diaphragm should be adjusted so that the light just fills the field of view. The higher the magnification of the objective lens, the smaller the field of view, so for the best image, the iris should be adjusted whenever you change the objective.

iris diaphragm

•••Practical tip•••

When you have the specimen in focus, look at your microscope from the side and see how far apart the objective lens and the slide are. It will always be this distance for that particular objective lens so remember what it looks like to make it easier to focus the specimen next time you use that objective.

•••Practical tip•••

If you can't focus, go back to the previous objective lens and focus with that first.

》Pointer

Objective lenses have a coloured band by which you can quickly identify them.

Objective magnification	Band colour
4	Red
10	Yellow
40	Pale blue
100	White

Magnification

The purpose of a microscope is to magnify small objects. In everyday speech, the word magnification means how much bigger an image is than the actual object. It's the same in microscopy, except that you can put numbers on it.

The object on the slide is first magnified by the objective lens. If all you had was an objective then the image would be four times, or ten times or forty times the object size, depending on which lens you use. The objective lens also determines the resolution of the image. The higher power the objective, the greater the resolution. This means that you can distinguish two points that are closer together, so a higher power objective allows you to see more detail.

Eyepiece lenses generally have ×10 or ×15 etched on the side so you can see the power of your eyepiece. The image from the objective lens is magnified by the eyepiece, so the total magnification of the image you see is the product of the two separate magnifications. The eyepiece magnification does not affect the resolution of the image. It just makes it bigger. We will consider magnification again when we look at biological drawings.

Calibrating the microscope

Calibrating a microscope lets you measure the actual size of structures on the slide. You need an eyepiece graticule, which is the ruler you can see. It is inside the eyepiece. It looks like this:

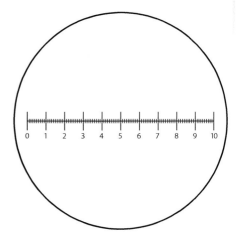

It looks the same through every objective because it is in the eyepiece. So with objectives of different magnifications, the divisions on the graticule represent different lengths. The higher the power of the objective, the smaller length each division of the graticule represents. So calibrating has to be done for each objective.

Grade boost

Remember how to define magnification:

$$\text{magnification} = \frac{\text{image size}}{\text{object size}}$$

Pointer

Calculation of magnification of image size with a ×40 objective and ×10 eyepiece:

magnification = objective × eyepiece

= 40 × 10

= 400

Grade boost

Remember how to define resolution:

resolution = smallest distance between two points that can be separately distinguished.

For a light microscope it is about 2 µm and for an electron microscope it is about 2 nm, although the microscopes in your school may not resolve quite such small distances.

To know what length each division represents, you need a stage micrometer. This is a microscope slide on which the object is a line 1mm long. It is ruled with markings for tenths and hundredths of a millimetre:

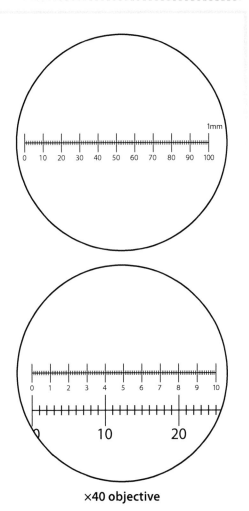

●●**Practical tip**●●●

Line up the zeros on the stage micrometer and the eyepiece graticule absolutely exactly.

Line up at the zero of the eyepiece graticule and the zero of the stage micrometer, making sure the scales are absolutely parallel. Look along the scales and see where they coincide again.

E.g. on a ×40 objective, 20 stage micrometer divisions may line up exactly with 80 eyepiece divisions:

×40 objective

Now three statements for the sums, remembering to say which objective you have used:

⩘ *Grade boost*

Remember to include a key for abbreviations:

smu = stage micrometer unit

epu = eyepiece unit

At a ×40 objective lens
1.   80 eyepiece units   =   20 stage micrometer units
           ∴1 epu   =   20/80 smu = 0.25 smu
2.   1 smu   =   0.01 mm
3.   1 epu   =   0.25 × 0.01 mm
                       =   0.0025 mm
Very small numbers of mm are better expressed as microns or micrometres (μm), so you can complete the calculation like this: =   0.0025 × 1000 μm
                                                                                    =   2.5 μm

●●**Practical tip**●●●

Many school microscopes have ×4, ×10 and ×40 objectives. The calibrations generally work out approximately as shown in this table.

Objective magnification	Length of 1 epu /μm
4	25
10	10
40	2.5

You can work out similar calibrations for all the objectives that you use. Microscopes differ slightly in their calibration so you have to do it again for each microscope you use. That is why, in school, it helps if you always select the same microscope. Then you only have to do it once. We will see how to use the calibration in measuring when we discuss biological drawings.

How to make microscope drawings

Making drawings of biological specimens is not art, although there is an art to it. The most important thing to think about is proportions. You are trying to show how structures look in relation to each other. The conventions used mean that when you see a biological drawing you can interpret it in the same way as the person who saw the specimen and made the drawing.

To start you need plain paper and a sharp HB pencil. The lines made by H or 2H are too pale and the lines made by B or 2B become too thick as your drawing progresses. A soft eraser will be useful while you are practising and developing your skill. Rest on several sheets of paper so that the surface pattern of your desk does not show through.

Support your elbow on the table and use it as a fulcrum so that your arm from the elbow down is free to move.

Your drawing must take at least half a page of A4 so practise making your lines long without taking your pencil off the paper. Lines should always look as if they have been drawn in one go.

Like this: **Not like this:**

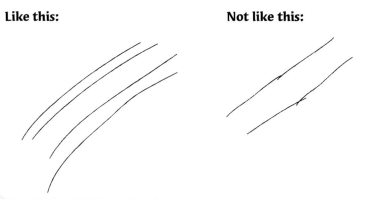

Boundary lines

In biological drawings, a line represents a boundary. In cells, this boundary is usually a membrane. So when you draw the boundary of a vacuole, for example, the line you draw is the tonoplast membrane, so you draw a single line which must meet. If the line does not meet exactly, it means there is a break, so circles must be completed.

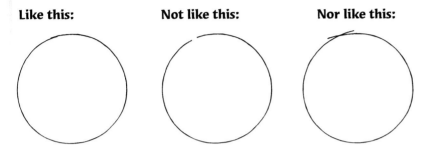

Like this: **Not like this:**   **Nor like this:**

If you draw anything thicker than a membrane it must have two boundary lines.

A cell wall is a thick structure so must always be shown with two lines.

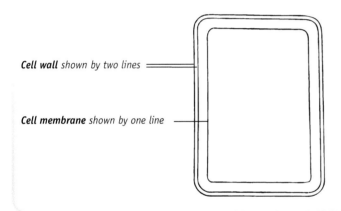

Cell wall shown by two lines

Cell membrane shown by one line

Shading

None. Cross-hatch if you feel you have to, but really it isn't necessary. A line showing a boundary is all you need to indicate the presence of a structure.

● ● ● **Practical tip** ● ● ●

Do not do any solid shading in your drawing.

Label lines

To make your label lines absolutely clear there are conventions to follow:

1. Lines parallel, as you can see in the drawing of the plant cell on page 11. Although if you have drawn something circular, such as a TS stem, you could use radial label lines:

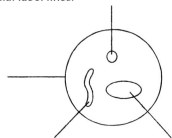

2. No arrow heads or dots at either end of the line
3. Always use a ruler
4. Your label line must end inside the structure you are labelling, not touching the outside and not hovering nearby:

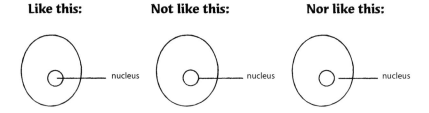

5. Don't let your lines cross:

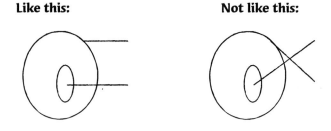

6. Make your label line long enough so that your label is well away from the drawing:

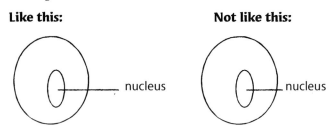

Boundaries

You may draw a sector of a circular structure, such as a TS alimentary canal. It is important to indicate that you have not drawn the whole structure and you can do that by ruled lines that define the edges of the part you have drawn. Sometimes it is described as looking like a pizza slice.

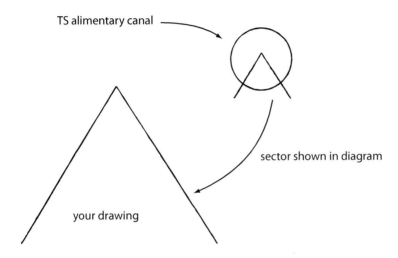

TS alimentary canal

sector shown in diagram

your drawing

If you draw a leaf section, you should include the midrib in your diagram. If, however, you are only drawing a portion of the lamina, you must include a small diagram to show where the section has been taken from, as shown here. The dotted lines indicate that what you have drawn continues beyond your diagram.

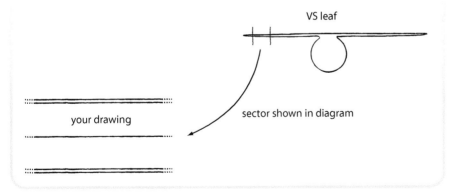

VS leaf

your drawing

sector shown in diagram

Drawing a low power plan

A low power plan shows layers of tissues but does not show individual cells so you are likely to use only the ×4 objective lens. It is a tissue plan, with lines showing where the tissues are and their relative sizes. Suitable specimens include VS leaf, TS artery or TS root.

If you draw a section through a leaf, most of your drawing will look like parallel lines at different distances from each other, although, depending on your specimen, you may see some vascular tissue to include. Similarly, if you draw a section through an artery, it may just appear to be concentric rings. The important point is that the proportions are correct.

<table>
<tr><td>Your leaf section will look like this:</td><td>Not like this:</td></tr>
</table>

• • **Practical tip** • • •

A low power plan shows only the layers of tissues, not individual cells. Only a high power drawing shows individual cells.

• • **Practical tip** • • •

The epidermis of a leaf and the endodermis of an artery are very thin compared with the mesophyll and muscular layer, respectively. They should still be drawn with two lines, as they have a significant thickness (the thickness of one cell, as opposed to the thickness of a membrane) but those lines must be very close together.

Grade boost

If your diagram has a feature with a precise position, make your measurements from there, e.g. the tip of a vascular bundle. That way, anyone looking at your diagram can see exactly where the measurements were made.

Grade boost

Somewhere on the page, have a key to say 'epu = eyepiece units'.

To show that your proportions are correct, you have to indicate the number of eyepiece units two of the tissue layers take. Rotate the eyepiece so that the graticule is aligned accurately for you to measure. In the leaf, it may be that the palisade layer is 1.2 epu thick and the spongy layer is 2.4 epu thick. Then you must make the spongy layer twice as thick in your drawing. You have to indicate where you have made the measurement and what the measurements are. Your examiner will check that the numbers you have written correspond with the thickness of the layers in the drawing.

There are two ways to write the actual numbers:

- Write 1.2 epu on the scale line itself.
- Put A at one end of the line and B at the other, then in a key somewhere else on the page write A–B = 1.2 epu.

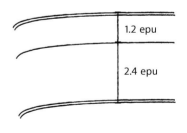

Put the lines to show where you have measured on the drawing itself, not to one side. Draw a little bar at each end, to show the limits of your measurement. Position it exactly.

Drawing at high power

A high power drawing of cells will be made with a ×40 or higher magnification objective lens. Choose a small number of cells, e.g. three. If a structure is thicker than a cell membrane, it must be bounded by two lines. Plant cells that are adjacent may share a cell wall and so your drawing must indicate this:

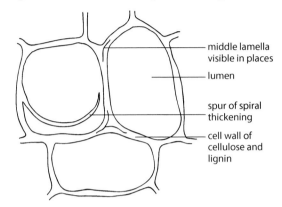

middle lamella visible in places

lumen

spur of spiral thickening

cell wall of cellulose and lignin

- Draw the inside of the cell walls first to get the correct shape and relative positions of the cells. You have automatically drawn the cell walls that they share.
- Then draw the cell walls around the outside, making sure you show that these outer cells share a cell wall with the three that you drew initially.
- Sometimes you can see a spur of spiral thickening – make sure it is continuous with the cell wall. A line means a boundary so take care to show that cell walls are continuous.

Remember that someone should be able to look down your microscope and identify the actual cells that you have drawn.

Showing the scale of your drawing

You must show the calibration for your particular objective lens somewhere on the sheet of paper. The simplest way of calculating it is shown in this book on page 9. Use that formulation to make sure your calculations are correct.

Having calibrated your microscope, you will know the actual length that the eyepiece units represent on the objective lens you have chosen. You can use this to calculate the actual length of what you have measured. For example, you may have a circular structure and have identified its diameter as a suitable measurement. Let us say that the diameter is 38 epu on a ×10 objective. Use the following steps to calculate its actual diameter:

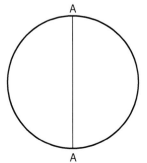

A

A

With a × 10 objective

A – A = 38 epu

1 epu = 10 μm

38 epu = 38 × 10 μm

∴ diameter = 380 μm

Grade boost

Go back and check that everything you have written is spelled correctly. If you're not sure, look it up.

Grade boost

As you are writing by hand, remember to underline the genus and species names.

••Practical tip•••

If you have more than a few hundred micrometres, it is sensible to convert it into mm, e.g.

$$380 \ \mu m = \frac{380}{1000} \ mm$$
$$= 0.38 \ mm$$

Pointer

Use a sensible number of decimal places. Your calculator may give you nine decimal places, but your drawing is unlikely to be correct to ±0.000000001 mm. If you are working in mm, 2 decimal places are enough. If you are working in micrometres, then one is enough.

Title

Diagrams must have a title. If you have made a slide with fresh material, you should write what the material is and write 'fresh specimen'. If you have used a ready-made slide, read the label and then copy what is relevant, e.g. VS *Ligustrum* leaf. Write 'prepared slide'.

Write down the magnification of the objective you have used, e.g. ×4 objective. This tells the person looking at your drawing how much detail it is reasonable to expect.

Units

The standard units for length are mm, but if your structures are small, it may be more suitable to use micrometres. The way you can decide is if you are working in mm but it starts 0.0..., then change to micrometres, so 0.034 mm is better expressed as 34 μm. But if, on the other hand, you have calculated that your structure is hundreds or thousands of micrometres, you should change to mm, e.g. 809 μm is better written as 0.81 mm (2 dp) and 1257 μm is better written as 1.26 mm (2 dp).

Labelling microscope drawings

Sometimes, it is not easy to match the structures you see in the microscope with what you have learned from your textbook. The more you look down a microscope, however, the easier it becomes to recognise what you are looking at. There are many useful books that show a photomicrograph and a drawing made from it, with labels on the important structures. Some of these are listed on page 73. If you use a book like this when you are examining specimens, you will learn how to identify structures. The list below tells you important components of biological specimens that you need to be able to recognise when you are doing microscopy, and these are the structures that would be suitable to label on your drawing.

Specimen	Suitable labels
TS artery	lumen, endothelium, tunica interna/intima, tunica media, tunica externa/adventitia
TS vein	labels as for artery
TS small intestine	columnar epithelium, mucosa/lamina propria, muscularis mucosa, submucosa, circular muscle, longitudinal muscle, serosa, villus
TS mammalian testis, e.g. rat, cat	spermatogonia, spermatids, area of spermatozoa, seminiferous tubule, Leydig cell, Sertoli cell
LS ovary, e.g. rat	germinal epithelium, primary follicle, primary oocyte, secondary follicle, Graafian follicle, secondary oocyte, zona pellucida, corona radiata, antrum, theca, cumulus cells, blood vessel, stroma, corpus luteum
TS spinal cord, e.g. rat	dura mater, pia mater, arachnoid, meninges (if membranes not labelled), white matter, grey matter, central canal, dorsal horn, dorsal root, dorsal root ganglion, ventral horn, ventral root, ventral median fissure
TS kidney, e.g. rat	Bowman's/renal capsule, glomerulus, capillaries, proximal convoluted tubule, distal convoluted tubule, basement membrane
VS leaf, e.g. *Ligustrum*, *Nymphaea*	epidermis, palisade mesophyll, spongy mesophyll, xylem, phloem, vascular tissue (if xylem or phloem not labelled), cuticle, collenchyma, stoma, air space, parenchyma, collenchyma, sclerenchyma
TS marram grass leaf (*Ammophila*)	cuticle, abaxial/lower epidermis, sclerenchyma fibres, palisade mesophyll/chlorenchyma, xylem, phloem, vascular bundle (if xylem or phloem not labelled), adaxial/upper epidermis
TS root	epidermis, cortex, endodermis, pericycle, xylem, phloem, vascular tissue (if xylem or phloem not labelled), parenchyma, root hair
TS stem	epidermis, cortex, medulla/pith, xylem, phloem, cambium, vascular bundle (if xylem or phloem not labelled), sclerenchyma, parenchyma, collenchyma
TS anther, e.g. *Lilium*	epidermis, tapetum/inner wall, pollen sac, xylem, phloem, vascular bundle (if xylem or phloem not labelled), area of dehiscence (if drawn), fibrous layer/outer wall, parenchyma

Grade boost

Use books such as those listed on page 73 when doing microscopy, so you can correctly identify what you are looking at.

●●●Practical tip●●●

In general, you should label a minimum of five structures accurately. Make sure your label lines are precisely placed, as described on page 12.

Drawings checklist

Item	Details	I can do this
Drawing	Sharp HB pencil	
	Paper resting on something while you draw	
	Drawing at least half a page	
	Accurate proportions	
	No individual cells in a low power plan	
	Lines meet exactly with no space and no visible join	
	Ruled boundary lines if drawing is a sector	
Label lines	At least five structures labelled	
	Label lines end in middle of structure or touching line, if membrane is labelled	
	Lines parallel or radial	
	Lines don't cross	
	No arrow heads or dots at ends of line	
	Lines drawn with a ruler	
	Labels away from the drawing	
Labels	All spellings correct	
	Genus and species names underlined	
Scale lines	Marked on drawing, not to one side	
	Little bars at the ends of line showing precise position of measurement	
	Measurement given	
Calibration	Three separate statements for your particular objective	
Measurement	Calculation of length of one indicated structure	
	Correct units given	
Key	epu = eyepiece units; smu = stage micrometer units	
Title	Drawing title	
	Preserved or fresh specimen	
	Objective magnification	

A.2 Aseptic culture

Culturing plant tissues and bacteria

At A Level, students often culture plant tissues or bacteria. Sterile, or aseptic, technique prevents the growth of fungi or unwanted bacteria. These contaminants may grow so fast that they obliterate the material you are trying to culture. Even if they are slow growing, they use up nutrients and may secrete metabolites that are toxic to your culture. Sterile, or aseptic, technique prevents the problem because:

- All the equipment has been sterilised, removing fungi, bacteria and their spores.
- All manipulations are performed in such a way that contaminants cannot enter the culture.

Sterilising equipment

Your school may have purchased apparatus already sterilised with gamma rays, e.g. Petri dishes, syringes and pipettes.

Glassware, dissecting equipment, white tiles, Petri dishes, syringes, pipettes and heat-stable solutions can all be sterilised in a domestic pressure cooker or in an autoclave. They are held for 15 minutes at 121 °C and high pressure, which kills bacteria, fungi and their spores. It is useful to seal the wrapping with autoclave tape before sterilising. The tape changes colour in the autoclave, so that you know the equipment inside the wrapping has been sterilised.

Inserting dissecting equipment into an autoclave.

If you are subculturing bacteria or dissecting plant material under sterile conditions, you should stand your equipment, e.g. wire loop, scalpel, forceps, in a conical flask containing a small volume of ethanol. Before touching the living material, the loop, scalpel or forceps are held in a flame to burn off the ethanol. They are heated until they glow and are maintained at red heat for a few seconds. They must cool before touching the culture material, so as not to kill it.

● ● ● **Practical tip** ● ● ●

Once you have opened a package that has been through the autoclave, you cannot guarantee its sterility. Do not reseal it for future use.

Setting up your work space

Do not work where there are draughts or air currents because they may carry contaminants, such as fungal spores. Your school may have a transfer chamber, which is an enclosed space in which you work but, if not, you can work on the lab bench. To avoid air movement, you must be sure that there are no open windows and that people are not going to walk by while you are working.

Before you start any sterile work, remove your watch and jewellery, tie back your hair, roll up your sleeves and wash your hands and wrists thoroughly.

The surface of the bench or transfer chamber must be swabbed with ethanol before you start. Let the ethanol evaporate away before you light the flame and begin work.

The flame and equipment should be arranged so that you do not have to reach over the culture while working, because contaminants might fall off your hand or arm on to the culture.

Setting up your cultures

A bottle or tube must be flamed before putting anything in or taking anything out. Flaming heats the air at the top of the bottle, which rises, moving any air and, more importantly, the contaminants it contains, up and away from the bottle opening.

Bacteria in liquid culture

Imagine you want to transfer a loop of bacteria from a solid culture in a universal bottle to a liquid medium in a stoppered culture tube. If you are right handed:

1. With your right hand, remove the wire loop from the ethanol flask and flame it, as described above. Hold it in the air to cool down for a few seconds. Do not put it down.
2. While it is cooling, take the bottle in your left hand.
3. With the little finger of your right hand unscrew and hold the lid. Do not put it down.
4. Pass the top of the bottle through a flame.
5. Take a sample of bacteria with the tip of the wire loop.
6. Re-flame the bottle and replace the lid.
7. Take the culture tube containing the liquid medium in your left hand.
8. Remove the lid with the little finger of your right hand and hold it. Do not put it down.
9. Flame the lid of the culture tube.
10. Shake the tip of the wire loop in the liquid medium.
11. Re-flame the tube and replace the lid.
12. Replace the wire loop in the ethanol.

Making a streak plate

Streak plating separates bacteria into individual colonies. This is useful if you want to isolate one species from a mixed culture or if you want to establish a new culture from a single bacterial cell. If you are right handed:

1. With your right hand, flame and cool a wire loop.
2. In your left hand, hold the container with the bacteria that you wish to separate. Remove the lid with the little finger of your right hand.
3. Flame the top of the container.

• • •Practical tip• • •

If you are left handed, the method is the same, but with the other hand.

Pointer

It takes practice to be able to manipulate your little finger independently of your thumb and first finger. Your sterile technique will improve.

4. Lift a bacterial sample out with the wire loop.

5. Re-flame the top of the container and replace the lid.

6. With your left hand, lift the lid of a Petri dish containing the solid agar medium you will be streaking. The lid should be lifted as small a distance as possible.

7. Make between three and ten streaks of bacteria in one quadrant of the Petri dish, taking care to touch the surface of the agar very gently, so that you do not make a hole.

8. Replace the lid of the Petri dish.

9. Re-flame the loop. You can test if it is cool enough by touching it on a part of the agar that has no bacteria. If the agar does not hiss or melt, the loop is cool enough.

10. Turn the Petri dish through 90°.

11. Place the tip of the wire loop in the streaked area and drag bacteria into a new quadrant, where you make another three to ten streaks.

12. Repeat steps 8–11 so you have a third set of streaks.

13. Replace the lid of the Petri dish.

14. Replace the wire loop in the ethanol.

15. Label the plate with the date and the species of bacteria. Place it, inverted, at 27°C for 48 hours, by which time the bacteria will have grown and individual colonies can be distinguished.

Streak plate after 48 hours

Culturing plant tissues

Plant material that is used as a source of material for culture, i.e. for explants, has bacteria and fungi on its surface. The plants are surface sterilised, so that the microbes do not contaminate the culture:

- Submerge the plant material in a solution of 10% (v/v) household bleach for 15 minutes.
- Wash the bleach off with three changes of sterile distilled water.

Suitable explants include:

- 1 mm slice of stem.
- 1 cm^2 piece of leaf tissue, with the lower epidermis peeled off.
- A shoot meristem.

To culture an explant, or to sub-culture an existing culture, the material has to be dissected. To be sure that it remains sterile:

- Dissect on a sterile white tile.
- Stand sterilised dissecting equipment in ethanol and flame it before using.
- Hold dissecting equipment as near to the end as possible, at an angle, to avoid contamination falling on to the material.

Plant tissues grow much more slowly than bacteria and fungi so, if any contamination is present, it has a far greater opportunity to grow. Sterile technique is, therefore, extremely important.

Arabidopsis thaliana, thale cress, growing in tissue culture.

>> Pointer

Make sure you know these five food tests: reducing sugars, non-reducing sugars, proteins, starch, and fats and oils.

>> Pointer

All sugars are white, crystalline, sweet-tasting and soluble in water. That is why we need tests to distinguish the different types.

Grade boost

Remember that when you describe the test for reducing sugars, you must write 'Benedict's reagent + heat'. Don't forget the heat.

>> Pointer

The Cu^+ is in red copper oxide, Cu_2O, which precipitates out. It appears as a different colour depending on its concentration and the proportion of Cu^{2+} remaining.

>> Pointer

When you write about the Benedict's test, please give it a capital B, as it was invented by Mr Stanley Benedict from Cincinnati. And give it an apostrophe before the s because it is his test. It is not the Benedictine test and has nothing to do with monks distilling alcohol.

A.3 Food tests

If you plan to be a biochemist, dietician or food technologist, one of your tasks may be to test foods to see what they are made of. You may have been doing this since Year 7 but at A Level it is much more interesting, even though the colour changes are still the same. There are five tests you have to know about and understand how to use. They are the tests for:

1. Reducing sugars
2. Non-reducing sugars
3. Proteins
4. Starch
5. Fats and oils

Test for reducing sugars

This was probably introduced to you in Year 7 as the test for simple sugars, but the simple sugars it tests for, like glucose and fructose, all do the chemical process of reduction. It is this that produces the familiar colour change from blue to brick-red.

You are sometimes asked to describe how to do the test and, as this question could be worth three marks, you have to make three separate points. They are:

1. Mix the test solution with an equal volume of Benedict's reagent.
2. Heat the mixture in a water bath at 70°C (or 80°C or 90°C or 100°C – it doesn't matter which as long as you state a temperature).
3. Observe the colour change from blue to orange (or red or 'brick-red').

Actually, when you do the test you will see that the colour the blue solution changes to depends on how concentrated the sugar solution is and how long you leave it. In fact, you could see a gradation from blue to murky green to orange to red to brown.

The colour comes from the reaction of the Cu^{2+} ion that gives the solution its blue colour. The sugar donates an electron to the copper ion and the ion is reduced to Cu^+. Donating an electron is a reduction so the sugar is a 'reducing sugar'. The reaction can be written like this:

$$Cu^{2+}_{(aq)} \quad + \quad e^- \quad \longrightarrow \quad Cu^+_{(s)}$$
$$\textit{blue} \qquad\quad \textit{from sugar} \qquad\qquad \textit{red}$$

The Benedict's test is sometimes described as being semi-quantitative. That means that you can tell which of various solutions has more or less reducing sugar in it and put them in relative order but you can't put a number on it. To do that, you would have to set up an experiment with different known glucose concentrations but keep everything else the same. Then you can use the final range of colours as a guide and match the colour given by the unknown concentration to a colour from a known concentration.

Safety note

Benedict's reagent is basic so it will damage your corneas badly if it gets in your eyes. Goggles are essential. It is an irritant if inhaled or ingested. If you get it on your skin, you must wash it off immediately. You will know because of the soapy feel that alkalis have.

Safety note

Avoid skin contact with hot water. If you are using an electrically heated water bath, take care not to splash around the electric socket. If you are heating your water over a Bunsen burner, be careful around the flame.

Test for non-reducing sugars

An examination question may ask about white, soluble crystals that do not give a red precipitate with Benedict's reagent. The question may ask how you would identify them. The first obvious thought is that they are sugar but health and safety concerns mean that you can't taste them. As they are not reducing sugar, the next logical step is to test for non-reducing sugar. This is what you do:

1. To 2 cm³ of the solution add two drops of the enzyme sucrase. Sucrase is supplied as a liquid, hence the units 'drops'.
2. Leave at room temperature for 5 minutes.
3. Add 2 cm³ Benedict's reagent.
4. Incubate at 70–100°C for 10 minutes.
5. If the blue of the Benedict's reagent has changed colour, a reducing sugar has been produced by the sucrase. This indicates that the sucrase had been acting on a non-reducing sugar. If the original non-reducing sugar had been sucrose, this is what has happened:

 Grade boost

When you describe the conversion of disaccharides into monosaccharides, say they are hydrolysed, not broken down. It's always best to use the correct technical term.

$$\underset{\substack{non\text{-}reducing \\ sugar}}{\text{sucrose} + \text{water}} \xrightarrow{\text{sucrase}} \underset{reducing\ sugars}{\text{glucose} + \text{fructose}} \xrightarrow{\text{Benedict's reagent / heat}} \text{red precipitate}$$

Some textbooks will tell you to hydrolyse the sucrose with hydrochloric acid by boiling the sucrose and acid together for 10 minutes. Then, in order to do the Benedict's test, you have to neutralise the acid with sodium hydrogen carbonate and keep testing with indicator paper until you achieve pH7. Using sucrase is quicker, more reliable, works at a lower temperature and allows you to deal with much smaller volumes of liquid.

But to return to the examination question, if you did not get a red precipitate after sucrose treatment, then you are not dealing with a sugar. The next step in the logic requires you to do the biuret test for molecules with peptide bonds, i.e. proteins and peptides.

Safety note

Sucrase is an enzyme and therefore made of protein. Some people have skin reactions to foreign proteins so if you get it on your skin, rinse it off immediately.

Something to try at home: drink some cold lemonade. It tastes sweet, because of the sucrose, and sharp, because of the citric and carbonic acids. Heat some and cool it again before tasting. Now it tastes like honey. This shows two biochemical points:

- Sucrose is broken down by heat into glucose and fructose, the main sugar in honey.
- Sucrose is hydrolysed by acid:

$$\text{sucrose} \xrightarrow{\textit{carbonic acid}} \text{glucose} + \text{fructose}$$

••• **Practical tip** •••

Holding the test tube at an angle allows the copper sulphate to trickle gently down the side of the test tube. This means it will form a layer on the top and not mix into the test solution.

Test for proteins

Make sure you can spell biuret. It is purple and made by a chemical reaction when you test for a protein. Do not confuse it with burette, which is long, thin and made of glass.

When you are asked to describe this test in an examination, it could be worth three marks, so make sure you give at least three separate statements:

1. Mix equal volumes (e.g. 2 cm³) of the solution you are testing with 1 mol dm⁻³ sodium hydroxide. You don't need to remember the concentration but if you give it, you are signalling to your examiner that you really know this.

2. Hold the test tube at an angle and gently trickle down the side of the test tube 10 drops 1% copper sulphate solution.

3. Where the copper sulphate meets the sodium hydroxide there is a reaction making blue copper hydroxide:

$$2\text{NaOH} + \text{CuSO}_4 \longrightarrow \underset{\textit{blue}}{\text{Cu(OH)}_2} + \text{Na}_2\text{SO}_4$$

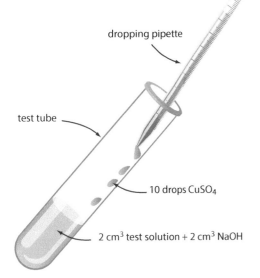

dropping pipette

test tube

10 drops CuSO₄

2 cm³ test solution + 2 cm³ NaOH

The copper hydroxide sits on the surface because you have trickled in the copper sulphate so gently, which is why this is sometimes called the 'blue ring test'. This happens whatever is in the test solution. It is the next step that proves you have peptide bonds present.

4. Cover the top of the test tube, invert it once and bring it back to a vertical position. If a protein is present, the blue colour will have disappeared and the solution will be purple. The purple compound is biuret.

The biuret test is another semi-quantitative test. The more protein there is, the darker the purple colour will be. If you have different shades of purple, you can put your test solutions in order of increasing concentration.

If you do the test with several solutions of known concentration, as long as all other factors stay the same, you can compare the purple they produce with the colour given by a solution of unknown concentration, and therefore, estimate the concentration of your unknown.

Test for starch

This is another test that you probably learned in Year 7. In those days you would have said that starch makes iodine go blue-black. It's still broadly true, but at A Level there is more to say about it.

First of all, at room temperature, iodine is a deep purple, crystalline solid. This test uses a solution of iodine in potassium iodide, so the important ion in this reaction is I_3^-. The I_3^- ion is made like this: $KI + I_2 \longrightarrow K^+ + I_3^-$. This is why we refer to 'iodine solution' rather than just 'iodine' at A Level.

The type of starch that this ion identifies is amylose, which is a helix of α-1,4 linked glucose molecules. The diagram shows how the I_3^- ion is held within the helix and it is the interaction of its electrons with the electrons of the amylose helix that produces the colour change:

The iodine solution is brown and the starch–iodide complex looks mauve, purple, blue-black or black, depending on its concentration.

Two factors affect this test:

- pH – low pH hydrolyses the starch, so the iodine test cannot be done in those conditions.
- Temperature – the amylose molecules and iodide ions have kinetic energy and are constantly moving. At room temperature, their movement is not enough to disrupt the complex. But if you put the test tube into a water bath to heat the complex, the kinetic energy of the I_3^- ions is enough to make it leave the starch helix and so the blue-black colour disappears. If you take the test tube out of the water bath again, you can probably guess what happens to the colour.

This test is often used to show the rate at which amylase digests starch. You will see that on page 44.

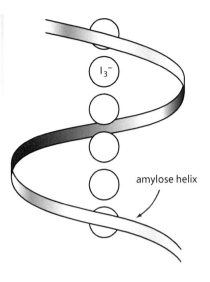

I_3^-

amylose helix

Test for fats and oils

1. Place a few drops of a liquid or a few mm³ of a solid sample in 2 cm³ ethanol and shake.
2. Leave for 2 minutes, while the lipid dissolves in the ethanol.
3. If using a solid sample, decant the ethanol into another test tube.
4. Add 2 cm³ of deionised water to the ethanol and shake.

If there are no lipids, no emulsion forms and the solution remains clear.

If lipids are present, a cloudy white emulsion forms at the top of the liquid. This is because lipids are soluble in ethanol and make a clear solution, but they are insoluble in water. When water is added to the ethanol, lipids come out of solution and form micelles, i.e. spherical groups of lipid molecules with their hydrophobic parts pointing inwards, away from the surrounding water molecules, and their hydrophilic groups pointing outwards, towards the water molecules. Lipids are less dense than water so the micelles float at the surface. They diffract light so they look cloudy.

Summary of tests

Learn these colour changes by heart as examination questions often ask you to state what they are.

Test for	Reagents	Initial colour	Colour of positive result
reducing sugars	Benedict's	blue	red
non-reducing sugars	hydrolysis + Benedict's	blue	red
proteins	sodium hydroxide and copper sulphate	blue ring	purple
starch	iodine solution	red-brown	blue-black
fats and oils	ethanol and water	colourless	cloudy white

Food tests checklist

Topic	Detail	I know this
All sugars	What are the physical properties of all sugars?	
Reducing sugars	Which test?	
	What reagents are needed?	
	How is the test done?	
	What is the colour change for a positive result?	
Non-reducing sugars	Which test?	
	What reagents are needed?	
	How is the test done?	
	What is the colour change for a positive result?	
Proteins	Which test?	
	Which bond does this test for?	
	What reagents are needed?	
	How is the test done?	
	What is the colour change for a positive result?	
Starch	Which test?	
	What reagents are needed?	
	How is the test done?	
	What is the colour change for a positive result?	
	What can you say about the temperature for the test to work?	
	What can you say about the pH for the test to work?	
Fats and oils	Which test?	
	What reagents are needed?	
	How is the test done for a solid sample?	
	How is the test done for a liquid sample?	
	What is the change for a positive result?	

A.4 Using whole organs and organisms

1. Use of animal dissection

There are three specified practicals requiring animal dissection: a fish head, a mammalian heart and a mammalian kidney. Some people say, 'I'm too squeamish' and don't like to even think about it. Even people who say they want to be a doctor or a dentist say that, and if you are one of those people, you may wish to consider another profession.

Some people say, 'I don't want to dissect because I'm a vegetarian'. You won't be eating it, so that argument doesn't work.

Some people say, 'It's cruel' but the reality is that the animals that are dissected in schools and universities are housed in clean, safe conditions where they have appropriate food and drink and live an infection- and predator-free environment. They live longer, healthier lives than would ever occur in the wild. If you are to dissect an organ rather than a whole animal, the organ would come from a butcher and so the animal from which it came is subject to all the laws governing the food industry.

Dissection is a wonderful way to learn. You can look at all the photos you like, read all the books, watch all the YouTube dissections and simulations you can find, but to do it yourself, you will learn the feel – the resilience of adipose tissue, the flexibility of blood vessels, the strength of the heart tendons and so on.

So take advantage when material is made available for you to dissect:

- You could use your phone to film your friend doing the dissection and provide a commentary on it.
- You could ask your teacher to show you more than is required by your specification. For example, in a fish head, the brain can be exposed and removed or an eye can be extracted. You could take out the lens, clean it and read through it.
- You may have the chance to dissect other organisms such as an earthworm, a locust or possibly larger animals such as a dogfish, frog, rat or rabbit.

Use every learning opportunity you can. When you are at an interview, showing enthusiasm for dissection will demonstrate that you are a committed student who has gained first-hand knowledge of how organisms are structured.

● ● ● Practical tip ● ● ●

If you like dissecting hearts or kidneys you can buy them at the butcher's and have a go at home. Try the liver as well. Go to the fish counter and buy a whole fish, such as a herring, to dissect its gut, gills and brain. Maybe, in the interests of education, the fishmonger will give you a fish head that the previous customer didn't want.

2. Whole plants, fruits and seeds

In the section of the specification about plant reproduction, you are required to know about wind- and insect-pollinated plants. The best way to support what you learn about this from the textbook is to examine flowers. Some people find plants boring, but that is usually because they haven't learned much about them. Just because they do not have scales, fangs or venom doesn't mean they can't wage war and protect themselves. Plants are fascinating.

Look at the flower structures so you really know what the 'small, green inconspicuous petals' of wind-pollinated flowers actually look like. Look at different types of flower so you can see anthers within insect-pollinated flowers, as opposed to anthers held outside wind-pollinated flowers by their long filaments. Actual flowers look different from models and diagrams, so make sure you look at real ones and that you look at as many different species as you can.

3. Use of models

When your teacher went to school, there were not many models available. Your teacher probably learned largely from textbooks. But you can relate what you learn in a textbook to a model which gives you the 3D version and so you understand so much better. You can see, for example, why the eye has a blind spot or why diffusion through the large air spaces in the spongy mesophyll of a leaf is efficient.

A model will help you learn a diagram, so if you are learning, for example, about the excretory system in mammals, improve your understanding of the diagrams by looking at a torso model and see how the components of the system are positioned in the body. See which kidney is higher.

Look at a flower model and understand the long distance the pollen tube has to grow. See the micropyle and understand that the pore remains when the integuments harden into the testa of the seed.

4. Use of preserved specimens

Your specification requires you to be able to distinguish the five kingdoms, based on the physical features of organisms. In some instances, it will be an advantage to know the differences in physical features between lower taxa, such as class, genus or species as well. Look at preserved specimens because it is these that will allow you to understand the significance of these features.

A.5 Looking at photographs

Biology is very visual so using photographs is a useful way of learning. A photograph may be used just to give you an image related to the subject matter, such as a portrait of Dolly the Sheep or a cactus. You may not have to use the photo in your answer but it helps you focus your thoughts.

Sometimes an examination question will ask you to provide evidence from a photograph, so you have to be careful that what you write actually refers to the picture, rather than what you assume is in it. You could be asked to give evidence from the photo of an advantage of fish farms. Now you can't see the individual fish so, even though you know it's true, you could not use 'no interspecific competition' as your answer, because you can't see which fish are there. But you could explain that having the fish farm closer to the shore than a fishing ground would reduce transport costs when the fish are caught.

You may be given a photomicrograph, i.e. a photograph taken down a microscope, or an electron micrograph, i.e. a photograph taken in an electron microscope. It could be a whole mount, i.e. a complete organism or it could be a thin section. If the electron micrograph were taken using a scanning electron microscope, you could be given a surface view. If the image is of cells, there are two ways you could be questioned. You may be asked to identify structures within a cell, so make sure you are used to looking at micrographs of cellular structures and that you can recognise them.

Or, given the magnification, you may be asked to calculate the length of one of the structures. The question will say 'show your working' so you must.

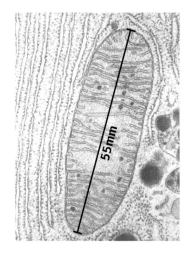

You may have an electron micrograph of a mitochondrion and be asked to calculate its actual length, given that it has been magnified 50,000 times.

Measure the mitochondrion's length. Let's say it is 55mm. As it has been magnified 50,000 times the image must be 50,000 times bigger than the real thing. So to find the actual size you divide the measured size by 50,000.

Always quote the formula:

$$\text{magnification} = \frac{\text{image size}}{\text{object size}}$$

$$\therefore \text{object size} = \frac{\text{image size}}{\text{magnification}}$$

Then substitute in the numbers

$$= 55 \div 50{,}000 \text{ mm}$$

$$= 0.0011 \text{ mm}$$

This is the answer in mm, which is correct, and you must, of course, include the units. But 0.0011 is a very small number and so a more suitable unit would be micrometres. So the last lines of your calculation would be:

$$= 0.0011 \times 1000 \text{ μm}$$

$$= 1.1 \text{ μm}$$

It might be useful to use the triangle method to help you get the equation right:

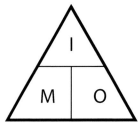

I = image size
O = object size
M = magnification

The triangle reminds you which way the equation works:

$$I = M \times O$$
$$M = I/O$$
$$O = I/M$$

Grade boost

Don't ever forget to put in the units.

When you have done a calculation, check that your answer is realistic and makes sense. In the example here, you might know that 1 μm is about right for a mitochondrion. But if your answer turned out to be, say, 250 μm, you would know that it was wrong, and so you would have to check your measuring and your working. Here are some sizes that are easy to remember.

Pointer

If you didn't know that mitochondria are about 1 μm long, you could work out if it was a suitable answer. Since Year 7 you have known that cells are often about 1/30mm, which is about 30 μm. You know how mitochondria look in relation to the whole cell, so that would tell you that 1 μm is a suitable answer.

Very approximate lengths:

Structure	Length /μm
Animal cells	10–30
Plant cells	10–100
Nucleus	10

Structure	Length /μm
Chloroplast	5
Mitochondrion	2
Bacterium	3

B: Experiments

B.1 How to design a lab experiment

When you study A Level Biology, you often think about how to plan and carry out experiments. Just as at GCSE there is likely to be an exam question that asks you to do that so it is important to know how to apply the basic principles.

General point about units

In Biology, the convention is to follow the criteria laid down by the Institute of Biology, so SI (Système Internationale) units are used. Here are the units that are used most:

	Length	Volume	Time
SI units	1 m = 1000 mm 1 mm = 1000 μm 1 μm = 1000 nm Notice how easy it is to convert – just × 1000 each time.	$1\ dm^3 = 1000\ cm^3$ $1\ cm^3 = 1000\ mm^3$	Seconds, abbreviated to s (note to Welsh speakers writing in English: that's 's' not 'e')
Note of caution	cm are not SI units so we rarely use them	l (litres) and ml (millilitres) are not SI units so we do not use them	Use minutes if the numbers of seconds are so large that the human brain cannot grasp them. Make sure you do not use a combination of minutes and seconds. Use only one unit at a time.

Grade boost

An exam question asking you to design an experiment may be worth several marks, so make sure you know this section thoroughly.

Pointer

The information given here will be useful to you whichever aspect of Biology you study because the same principles apply.

Experimental design

There are five major areas to consider: variables, control experiment, accuracy, repeatability and the numerical analysis. Of course, these sections all have sub-sections. Here they are:

1. Variables

A variable is something that can change. It may be you, the experimenter, who changes it, it may be the different readings you take and it may be something in the surroundings which has no bearing on your experiment, but may change anyway and then make a difference. We will look at three types of variable.

a) **Independent variable** – this is the factor you choose to change. You decide its values so they are fixed and do not depend on anything else that is going on. That is why they are 'independent'.

> **Example:** In an experiment to test the effect of pH on the reaction rate of the hydrolysis of hydrogen peroxide by catalase, you would measure the rate of reaction at several different pH values that you have chosen.
>
> So pH is the independent variable.

At A Level, when designing an experiment, use a minimum of five values for the independent variable. One of the values could be zero. In that case, you would add water instead of your test solution, to keep the total volume, and therefore the concentration of everything else, the same.

b) **Dependent variable** – this is what you plot on your graph. It is what you count or measure or what you calculate from your counts or measurements. The dependent variable depends on the independent variable, hence the name.

> **Example:** In an experiment to test the effect of carbon dioxide concentration on the rate of photosynthesis of pond weed, you would measure the rate of photosynthesis for several different concentrations of carbon dioxide.
>
> So rate of photosynthesis is the dependent variable.

c) **Controlled variables** – in an experiment, the only thing you want to change to get your readings is the independent variable. Everything else about your experiment stays the same so you have to control them.

> **Examples:** If the temperature kept changing in an enzyme experiment, or the light intensity kept changing in a photosynthesis experiment, your results would not be valid. Such factors must be kept constant so they do not affect your results. They are therefore controlled.

When you do enzyme experiments, there are several factors that can easily be changed. They are:

- enzyme concentration
- enzyme volume

> ≫ **Pointer**
> Make sure you distinguish between volume and concentration. If you have 2 cm^3 of 1% enzyme, it is twice the volume of 1 cm^3 of 1% enzyme. Its concentration is the same.
> If you have 2 cm^3 of 2% enzyme, it is twice the concentration of 2 cm^3 1% enzyme, but its volume is the same.
> Students sometimes think that 2 cm^3 is twice the concentration of 1 cm^3 enzyme but you will not make that mistake.

- substrate concentration
- substrate volume
- temperature
- pH.

In any enzyme experiment, one of these will be the independent variable. All the others must stay constant, so if you can remember this list, it will always be easy to think about controlled variables.

>> *Pointer*

Notice the spelling – you choose some variables and keep them the same. You control them and keep them constant so they are 'controlled variables'. The 'control experiment' is something else entirely.

2. Control experiment

When you do an experiment and get your results, your meanest critic might say, 'But that's going to happen anyway. It's nothing to do with your independent variable.' To prove them wrong, you set up a 'control'. This means you take away the most significant element in the experiment. Then, when you run the experiment, you get no change in the results. In this way, you can see that the change in the results (dependent variable) is directly because of the change in the independent variable.

> **Example:** In an experiment to test the effect of substrate concentration on the rate of hydrolysis of starch by amylase, denature the amylase by boiling for ten minutes then cool it before use. It is denatured and cannot catalyse a reaction so there is no digestion.
>
> This shows that when you had different results before boiling, it was because of the enzyme, so your results showed the true effects of starch concentration.

If your independent variable is a concentration, don't use zero concentration as a control, because zero is just another value for the independent variable. Denaturing the enzyme is generally the most suitable control.

If you are investigating photosynthesis or transpiration, it is tempting to think about zero light intensity as a control but then you could not see to do your experiment. Also, if the independent variable is light intensity, zero is just another value. So for these types of experiments, it makes sense, when you have collected all your results, to do it one more time with all the leaves removed as the control.

>> *Pointer*

The control experiment removes a vital agent of your experiment, proving that it is the independent variable that is responsible for the difference in your results.

>> *Pointer*

Do not use a zero value of your independent variable for the control experiment. Zero is just another value for the independent variable.

3. Accuracy

At A Level, you know how to use the apparatus properly so we have to assume there is no human error. Only you know if that is justified. This means that the accuracy of your readings depends on the accuracy of your equipment. We can think about this is two ways:

- **Objective readings** are made when you read off a scale and anyone making the reading will get the same answer, e.g. volume in a measuring cylinder, time on a stop-clock. You are limited by how far apart the graduations are.

>> *Pointer*

'Accuracy' refers to how true your readings are.

Grade boost

Make sure you know which of your readings are subjective and which are objective.

> **Examples:** If you measure volume in a 10 cm³ syringe graduated in cm³, you can, in theory, estimate to the nearest 0.1 cm³, but you are only accurate to the nearest cm³, i.e. you have an accuracy of ±0.5 cm³.
>
> If you measure length with a 15 cm ruler graduated in mm, you can, in theory, estimate to the nearest 0.1 mm, but you are only accurate to the nearest mm, i.e. you have an accuracy of ±0.5 mm.

- **Subjective readings** have an element of your personal judgement, such as when you match a colour to an indicator paper, or decide when a solution has changed colour and what colour it has changed from and to. Everyone's eyes are different and people judge in different ways, so it may be that no two people would agree. So here, the subjective judgement is a limitation on the accuracy of the experiment.

4. Repeatability

Pointer

Repeatability, reliability and consistency refer to how close your repeat readings are to each other. You might even hear the word replicability, but probably not from an A Level student.

This is sometimes called reliability or consistency and it describes how likely you are to get the same results more than once. If you do an experiment once, you may get an accurate result or it may be a bit off or it may be completely wrong. You have no way of knowing which. If you use your apparatus properly and take readings with great care then every time you do the experiment you ought to get the same result. But it doesn't happen like that in real life, because very small differences in the set-up or timing make a difference to the result, as does innate biological variation. In fact, if you get exactly the same results when you do your repeats, it looks rather suspicious.

So to get around this problem, it is normal to take readings several times. Three times is suitable in an A Level experiment. Then you can calculate a mean result. The mean is more reliable than any individual result because it takes into account all the slight variations that have happened and all the individual readings are scattered around it. The more repeats you use to calculate the mean, the more reliable the mean is. Every time you take a reading, it is as accurate as every other time, but the mean is more reliable than an individual reading because it takes all them into account.

Grade boost

The mean is no more accurate than an individual reading but it is more reliable.

Another advantage of using many results to calculate a mean is that if you get a fluke result that is clearly very different from all the others, you are entitled to discard it in your calculation. That isn't cheating. It is a way of acknowledging that in Biology systems are very difficult to replicate exactly.

Pointer

What's the difference between reliability and reproducibility?

Reliability refers to the consistency of the results.

Reproducibility refers to doing an experiment another way and getting the same results.

Reproducibility can give you confidence in the reliability of your results but they are not the same thing.

What's the difference between accuracy and reliability?

Think of it this way: every morning you get on your bathroom scales and you weigh 50 kg. You eat healthily and exercise the same amount each day so you don't expect your weight to change. And you weigh 50 kg every morning. Your scales are very reliable because you always get the same result. But little known to you, your evil twin has adjusted the scales so that every time, they weigh 5 kg too little. So although they are highly reliable, they are not accurate.

Reliability is about getting the same answer each time. Accuracy is about getting the true answer.

These are not the same thing.

5. Numerical analysis

A. Decimal places (dp)

The number of decimal places you use depends on how the apparatus is graduated. If you are using a data logger, then you can choose how many dp you read to, but choose sensibly. One may well be enough. If you are using a scale graduated in mm then you can read to the nearest mm and record in whole numbers of mm.

If you are timing in seconds, only read to the nearest second. Your response time and the limitations of your equipment mean that this is the most accurate you can be. So no dp in this case.

If you are calculating a mean from repeat readings, it is acceptable to quote one dp more than you have given in your results table. But don't forget to always give the same number of dp, even if you calculate to a whole number, you must still quote the same number, e.g. 3.0, not just 3 if all the other means are given to one dp.

B. Rate of enzyme-controlled reaction

There are a lot of experiments, especially at AS/Year 1 of your Biology course, which demonstrate the properties of enzymes. It is useful to make calculations from your data that show the rate of reaction. The way you do this will depend on the particular experiment.

> **Example 1**
> **Investigating the rate of hydrolysis of hydrogen peroxide by catalase**
> You collect oxygen by displacement of water and read its volume in either an inverted measuring cylinder or a gas syringe. The faster the reaction is happening, the more gas you collect in a given time, so the volume of oxygen represents the rate of reaction and no further calculations are needed.

> **Example 2**
> **Investigating the rate of digestion of casein in milk by trypsin**
> You time how long it takes for milk to clear. This may be done subjectively, by eye, or by having the reaction take place in a test tube in a colorimeter. Then you can time how long it takes for the milk to reach a pre-decided absorbance. Either way you will measure a time. To convert that to a rate, you find the reciprocal because rate = 1/time.

When you analyse your experiment, remember to discuss either time or rate, depending on what your aim is, but don't get the two confused.

> **Example 3**
> **Measuring the rate of respiration in yeast**
> Your experiment may be to measure the rate of respiration in yeast, in which case you may be measuring the time an indicator takes to decolorise. Respiration is a sequence of enzyme-controlled reactions and so this experiment can be thought of in the same way as the enzyme experiments. In this case, the rate is given by rate = 1/time. But remember to use either minutes or seconds and not a combination of the two.

Grade boost

Always use the same number of dp for your results and indicate how many in the column heading.

Pointer

Calculating rate of reaction depends on the experiment:

Examples 1 and 4:
rate of reaction = volume

Examples 2 and 3:
rate of reaction = 1/time

When you measure the rate of photosynthesis, you are also measuring the rate of a sequence of enzyme-controlled reactions.

Example 4

Measuring the rate of photosynthesis of *Elodea*, Canadian pond weed

In this experiment you may measure the volume of oxygen produced by a given mass of *Elodea* in a given time. The volume collected is a measure of the rate of photosynthesis but you would quote it as a volume (cm^3) and explain in your analysis why this is equivalent to the rate of photosynthesis.

C. Standard notation

If you are calculating a rate of reaction using a reciprocal you may have zeros after the decimal point. It is perfectly acceptable then to multiply all your results, e.g. by 100 to remove the zeros and have one number before the decimal point and $\times 10^2$. This is called standard notation. It's much easier to plot. Don't forget to indicate that you have done this in the column headings.

Example

An experiment testing the effect of pH on the rate of production of a colour change

pH	time for indicator to change colour /s	1/time /s^{-1} (4 dp)	1/time $\times 10^2$ / $s^{-1} \times 10^2$ (2 dp)
3	120	0.0083	0.83
5	80	0.0125	1.25
7	40	0.0250	2.50
9	90	0.0111	1.11
11	140	0.0071	0.71

Same results *Much easier to plot*

D. Percentage change

Any conclusion you draw will be enhanced if you can quote a percentage change. This is Year 8 arithmetic so should not frighten off a sixth form student.

Calculate it like this:

$$\% \text{ increase} = \frac{\text{actual increase}}{\text{initial value}} \times 100\% = \frac{\text{final value} - \text{initial value}}{\text{initial value}} \times 100\%$$

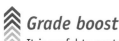

Grade boost

It is useful to quote a percentage change in the conclusion of an experiment.

E. Q_{10}

If you investigate the effect of temperature on the rate of an enzyme-controlled reaction, you will get results that allow you to plot a graph like that shown on page 37. One way of assessing your accuracy is to calculate Q_{10} from your graph, because theory tells us it should equal about 2. The closer your calculated value is to 2, the more accurate it is.

≫ Pointer

The closer your calculated value of Q_{10} is to 2, the more accurate your experiment has been.

Draw a line of best fit using only the points where the rate is increasing.

Then read the rate of reaction at two temperatures that are 10°C apart.

On the graph
rate at 30°C = 22 AU
and rate at 20°C = 15 AU

$$Q_{10} = \frac{\text{rate at } (t+10)°C}{\text{rate at } t°C} = \frac{22}{15} = 1.47 \text{ (2 dp)}$$

That means that the set of results that produced this graph is not very accurate, but yours might be better. It doesn't matter as long as you can discuss the sources of error and how you might improve the experiment.

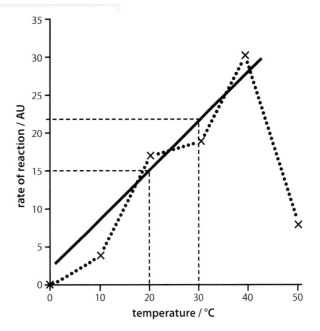

F. R_f

When you learn about photosynthesis, you may separate photosynthetic pigments by chromatography. You will end up with a plate or paper strip with pigment spots at various distances from the origin. Each pigment travels a characteristic proportion of the total distance moved by the solvent. If you work out this proportion, called R_f, you can identify each pigment.

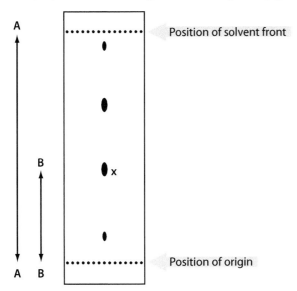

For the pigment spot labelled x, $R_f = \dfrac{\text{distance moved by x}}{\text{distance moved by solvent front}} = \dfrac{B}{A}$

You can measure the distances with a ruler and calculate R_f for each spot and identify them all.

B.2 How to design a field work experiment

1. Why do field work?

When biologists talk about field work, they are referring to looking at plants and animals living and growing in the wild, not in a lab. In a lab, you can control environmental factors like temperature and light intensity, but in the wild, you can't. So field work is a way of looking at plants and animals in their natural environment and seeing how they behave in the real world.

Not only can you design experiments to ask questions about how individuals behave, but also about how groups of individuals, whole populations of one species or communities of many species behave. You could test the effects of physical factors, such as the effect of soil pH on heather population density, or the effect of water depth on water shrimp population. You could test chemical factors in the environment, such as the effect of nitrate concentration on the number of bloodworms in a stream.

Field work techniques are powerful ways of monitoring populations and the effects on them.

At A Level, you are not required to do specific field work exercises but if you do, you will understand parts of your specification much better. You will also have a wonderful time.

>> **Pointer**

In these days of climate change, pollution and soil degradation, we will need more ecologists to find out what is happening and publicise the threats to species, populations, communities and their habitats. This could be you.

2. Designing field work experiments

The same principles apply to field work as to lab-based experiments so you have to consider variables, accuracy and reliability in the same way. But there are particular issues you have to think about because, unlike in a lab, you cannot control significant environmental factors. And because organisms show variation, to understand properly what your results mean, you have to interrogate the data with statistical tests. One thing at a time.

A. Variables

Your field work experiment will test the effect of an environmental factor, the independent variable, on something biological, the dependent variable. Here are some examples:

Independent variable	Dependent variable
Light intensity	Area of bramble leaves
	Length of soft rush leaves
Soil compaction	Percentage cover of plantain
Distance from drainage outlet into a stream	Number of bloodworms
Water depth in a stream	Number of water shrimp
Water flow rate in a stream	Number of mayfly larvae
Outcrop aspect	Species diversity of lichens
Soil water content	Percentage cover of moss

B. Controlled variables

As in a lab experiment, you want only the independent variable to change, but in the field you cannot control the environment. So you have to choose the sites for your experiment which are as similar as possible in every way other than the independent variable. Then you would have to monitor the most important factors, to ensure that they stay similar enough not to make a significant difference to your dependent variable.

> **Example 1**
>
> If you are doing an experiment with plants, among the most important factors in their growth are the light intensity and air temperature. So if you were investigating the effect of mowing on the species diversity of plants, you would choose a long grass area and a short grass area but do your best to ensure that they have the same light intensity, by making sure that the tree cover was similar and that they have the same air temperature by doing the whole experiment in as short a time as possible.

> **Example 2**
>
> If you are doing an experiment with aquatic invertebrates, such as the effect of nitrate concentration on leech numbers, you would only sample in areas that had similar water depths and flow rates.

C. Sampling

i) If your experiment is along an environmental gradient, you would use a transect. This is a line along which you sample. The further you go along the line, the further along the gradient you are.

> **Example**
>
> Investigating the density of limpets at increasing height above sea level. The further along the transect you go, the higher above sea level you are. You would sample by placing a quadrat at regular intervals along the line and count how many limpets in each.

The actual length of your transect depends on where you are. It needs to be long enough to cover enough of the gradient to give you a meaningful result. But if it is too long, at the ends, there won't be any change, and so there is no need to take results there. It needs to be just long enough to cover the whole change.

ii) If there were no environmental gradient, but a uniform environment, in theory you would count every shrimp in the river or measure every ground ivy leaf in the field. Obviously you can't, so you have to sample. You are after enough organisms to get a reliable mean and they have to be representative of the whole population.

• • • Practical tip • • •

A transect may be 10 m long, going from a wooded area into an unshaded area, but it could be 50 m long if you are going down a hill looking at the effect of soil water content.

≫ Pointer

Sampling is the selection of individuals from a population which allow you to estimate the characteristics of the whole population.

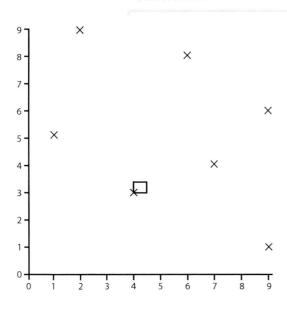

The way to do it is called block sampling. You set up two axes, at right angles and use random coordinates at which to sample. Let's say you want 15 samples. Go up to the first 15 people you see and ask them for the last two digits in their phone number. Here are your coordinates. The first of the digits is your x coordinate and the second is the y coordinate.

If you are placing a quadrat at each coordinate, always put the same corner of the quadrat on the coordinate. People usually use the bottom left-hand corner, as shown in the diagram. This is a way of being consistent and enhances your overall reliability.

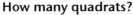

How many quadrats?

You can decide beforehand how many quadrats to do. But if you are not sure, you could calculate a running mean. This means that you calculate a new mean every time you do a quadrat reading. When the mean is the same for two quadrats in a row, you have enough. Do just one more.

Grade boost

Using this way of checking the number of quadrats would be a useful improvement to write about in the analysis of your experiment.

Type of quadrat

i) If you are looking at the distribution of the alga *Pleurococcus* on tree trunks, it is best to use a grid printed on to acetate, which you can bend around the trunk and see through.

ii) If you are investigating water beetles in a 5 m wide stream, a solid, open frame quadrat is best.

iii) If you are investigating percentage frequency of plants along a transect, you could use a 0.5 m point frame quadrat.

iv) If you are assessing percentage cover, a gridded quadrat is best.

But what size quadrat?

Often, you have no choice, because the standard 0.5 m × 0.5 m is all that is available. You can work out the best size by a preliminary experiment. Start with a small quadrat and make the counts. Then double the area and repeat the counts. Keep doubling the area and repeating the counts until doubling the area does not increase the number you count. Then you have found the optimum area of quadrat.

Grade boost

Using this way of checking the right size quadrat would be a useful improvement to write about in the analysis of your experiment.

D. Control experiment

You can't do a control experiment in the sense that you inactivate the independent variable. However, just as in a control experiment in the lab, you must keep all the variables constant, other than the independent variable that you have chosen. This is not always possible because, in fact, you cannot control the conditions in the field. So you choose your sampling areas to be as similar as possible, with the exception of the independent variable.

E. Reliability

As with lab experiments, you have to take multiple readings and calculate a mean. Each quadrat position gives you a reading and with block sampling, all you do is average those you have taken. But along a transect you only have one point at each distance. You could put the quadrat either side of the line to give two readings at each distance and calculate a mean of those or, depending on the situation, you could lay three transects parallel with each other and close by. Then you take results at the same distances along each. These are the replicates.

F. Accuracy

In your experiment, you will be measuring the effect of an abiotic factor on animals or plants. The measurements of the abiotic factors, such as concentration of dissolved oxygen in the stream, soil temperature or light intensity in the field are likely to be accurate because you will use properly calibrated equipment. It is the biotic data which limits the accuracy of the experiment overall and this will vary enormously, depending on your experiment. You may be counting numbers of larvae and some may be hidden in weeds, making it impossible to see them. Or you may be identifying plants, and without considerable experience, they can be misidentified. So you do the best you can and admit the problems. You can shake the weeds in a water sample and get out as many larvae as you can. You can take a field guide to help you identify plants.

G. Risk assessment

The major hazards in field work are very different from those in the lab. Whatever habitat you are in, sunburn is a hazard. In a terrestrial habitat, bites and stings are hazards, especially if you have particular allergies. Slips and trips and falling branches also have to be considered. If you are working in an aquatic habitat, Weil's disease, caused by the bacterium *Leptospira*, must be avoided and so you cover all cuts and wear rubber gloves or surgical gloves to protect yourself. Drowning is best avoided.

3. Graphs

The type of graph you draw will depend on the experiment you have done. There are four common ways of displaying field work data:

- **Line graph** – if you are investigating the number of bloodworms at different distances from a sewage outlet into a stream.
- **Scatter graph** – if you are testing the effect of light intensity on the width of ground ivy leaves.
- **Bar charts** – these are useful if you are comparing two habitats and have calculated the mean value of your biotic factor for each, e.g. if you have investigated the percentage cover of lichen on north and south facing rocks.
- **Kite diagrams** – these are used if you are looking at the changing distribution of plants along a transect. They show the changes very clearly.

>> *Pointer*

If you are testing a correlation, repeat readings are not possible. You must take enough readings to make your statistical test valid. This is discussed on page 42.

 Grade boost

The least accurate part of a field experiment is usually to do with assessing the dependent variable, i.e. the biotic factor. It's worth explaining that in your write up.

4. Analysis of field work data

Sometimes the aim of your experiment is easily answered:

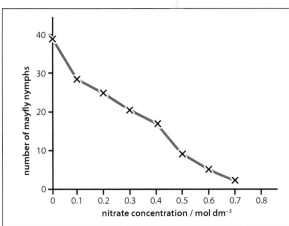

Example

Aim: to test the effect of nitrate ion concentration on the number of mayfly nymphs in a stream.

You would clearly see that the number of mayfly nymphs decreases as the nitrate ion concentration increases. This is because the increase in nitrate concentration is correlated with a decrease in concentration of dissolved oxygen. Mayfly nymphs have a high dependence on dissolved oxygen and their numbers fall significantly as the oxygen concentration decreases.

For an experiment like this, you could show your results with a graph as shown here.

A statistical test could be used to test if there is a correlation, but if your results are very clear, at A Level, you might choose not to. When you become a politically aware ecologist, however, statistics would be essential evidence for your cause.

Sometimes the data do not give you such a clear result and then you would have to use statistics to test the meaning of your results. You must state a null hypothesis, as this hypothesis is what you are testing.

 Pointer

The null hypothesis states that the independent variable has no effect on the dependent variable. You must always give a null hypothesis at the start of a statistical test.

Example

Aim: to test if there is a difference in the height:width ratio of periwinkles on rocky shores with high or low exposure.

Null hypothesis: there is no difference in the height:width ratio of periwinkles on rocky shores with high or low exposure.

Your experiment and statistics test if the null hypothesis can be rejected. It is a philosophical matter to seek to reject the null hypothesis, rather than accept it, because it is easier in experimental work to show a suggestion is untrue than to show that it is true.

 Grade boost

Note the terminology – statistics do not prove or disprove. It tests a null hypothesis and then either rejects or accepts it.

Which test?

There are three tests that are especially useful in this sort of work and it is easy to decide which to use. Decide before you start the experiment which you expect to use. This may have a bearing on the number of readings you make.

1. Use the t-test if you are comparing the means of the two sets of data. Three requirements make this test valid:
   - Each set of data is normally distributed – plot a frequency histogram to see. An approximately normal distribution means you can use this test.
   AND
   - you have at least 15 samples in each data set
   AND
   - you are using raw data, not manipulated data.

Pointer

There are three common statistical tests for field work. Make sure you choose the right one: Student's t-test, Mann–Whitney U-test, Spearman Rank Correlation test.

2. Use the Mann–Whitney U-test to compare two medians. You can use this test if:
   - the data is not normally distributed, which you can tell from the frequency histogram

   OR

   - you have fewer than 15 samples in each data set

   OR

   - the data is manipulated.

» Pointer

Please don't call it the Mann-U Whitney test.

3. Use the Spearman Rank Correlation test if you are looking for a correlation:

> **Example**
>
> **Aim:** to test the effect of light intensity on the percentage cover of meadow grass.
>
> **Null hypothesis:** there is no correlation between the height of meadow grass and the light intensity at which it grows.

Make sure you have at least 15 sets of results. If you plot a graph of your results, you do not need to add line of best fit. The statistics will tell you if a line is justified or not.

Experiment design checklist

Design area	Detail	I can do this
Aim	Clearly state aim in terms of two variables	
Prediction	Describe what you think will happen, without explanation	
Independent variable	Clear statement	
	Five values	
Dependent variable	Clear statement	
	Units	
Controlled variables	Decide the two most important and name them	
	Values for each, with units	
Repeats	State how many	
Control experiment	Identify which feature is inactivated	
	Identify what stays the same	
	If there is no suitable control, explain why not and state what control is for	
Hazard	Identify hazard	
	Know what the risk is	
	Know how to minimise risk	
	Know what to do if hazard occurs	
Construct results table	Detailed column headings	
	Units in column headings only	
	Enough repeats or an explanation as to why not	
	Appropriate data recording to suitable number of decimal places	

B.3 Standard experiments

This section will introduce you to some standard experiments and give examples of how to approach their design. You may do similar experiments and so you can adapt this information to suit what you are doing in the lab or in the field.

Enzyme experiments

Aim	Independent variable	Dependent variable	Method summary	Controlled variables Variable	Controlled variables Value	Control
To show the effect of temperature on rate of reaction of amylase digesting starch	Temperature, e.g. 20°C, 40°C, 60°C, 80°C, 100°C	Rate of digestion = 1/time taken for complete starch digestion / min^{-1} or s^{-1}	Time incubation period required for all the starch to be digested so that iodine solution does not turn black when added	Enzyme concentration	0.1 g/100 cm³	Boil the amylase for 10 minutes and cool it for the experiment at each temperature
				Starch concentration	1.0 g/100 cm³	
To show the effect of pH on the volume of oxygen produced when catalase digests hydrogen peroxide	pH, e.g. 2, 4, 6, 8, 10	Volume of oxygen produced in 5 minutes/cm³	Put potato discs in different pH buffers. Measure oxygen volume produced by their catalase 5 minutes after the addition of hydrogen peroxide	Concentration of catalase by having the same number of potato discs	30 discs	Boil the potato discs in water for 10 minutes and cool before using at each pH
				Hydrogen peroxide concentration	20 vol	
To show the effect of casein concentration on the rate at which trypsin digests it	Casein concentration provided by different concentrations of milk powder, e.g. 0, 1, 2, 3, 4 g/100 cm³	Rate of digestion = 1/time for milk to change from white to clear / min^{-1} or s^{-1}	Time how long each mixture of milk and trypsin takes to clear	Temperature	30°C	Boil the trypsin for 10 minutes and cool then use it at each concentration
				Trypsin concentration	0.5 g/100 cm³	
To show the effect of lactase concentration on the rate of lactose digestion using immobilised lactase	Lactase concentration, given by different number of immobilised lactase balls, e.g. 0, 20, 40, 60, 80 balls	Concentration of glucose produced /mol dm⁻³	Incubate same volume of milk over different numbers of lactase balls for same time. Use glucose-measuring strips to measure glucose concentration in product	Temperature	25°C	Boil the lactase for 10 minutes and cool before making balls; perform the experiment at each ball number
				Milk concentration	4 g/100 cm³ milk powder	

Photosynthesis experiments

Aim	Independent variable	Dependent variable	Method summary	Controlled variables Variable	Value	Control
To compare R_f values for pigments in spinach leaves	Pigment type, e.g. a. chlorophyll a and b. carotene, xanthophyll	R_f = distance moved by pigment/distance moved by solvent front	Separate spinach pigments by paper chromatography with propanol/petroleum ether solvent	Temperature / Propanol:pet ether ratio	25°C / 4:1	n/a
To determine the effect of light intensity on the rate of photosynthesis given by different distances of light source	Light intensity given by different distances of light source, e.g. 0.2, 0.4, 0.6, 0.8, 1.0 m	Time for hydrogen carbonate indicator to turn from red to yellow / min or s	Suspend algae, e.g. *Scenedesmus* in alginate balls and incubate in hydrogen carbonate indicator at various distances from light source. Time the indicator's colour change from red to yellow	Temperature / Carbon dioxide concentration	25°C / 0.05 mol dm^{-3} NaHCO3	Boil algal suspension for 10 minutes and cool before making alginate balls

Transpiration experiments

Aim	Independent variable	Dependent variable	Method summary	Controlled variables Variable	Value	Control
To determine the effect of light intensity on the rate of transpiration	Light intensity given by light distance from shoot source, e.g. 0.2, 0.4, 0.6, 0.8, 1.0 m	Transpiration rate, measured by the distance travelled by a meniscus along a capillary tube /mm	Set up potometer and take readings of meniscus movement with light source at different distances along a metre rule. Equilibrate shoot for 5 minutes with lamp at each distance before making readings	Air movement / Air humidity	Still air / Ambient humidity	Repeat experiment with all leaves removed
To determine the effect of air movement on the rate of transpiration	Air movement given by settings on a cold air hair dryer, e.g. 0, 1, 2, 3, 4	Transpiration rate, measured by the distance travelled by a meniscus along a capillary tube /mm	Set up potometer and take readings of meniscus movement with hair dryer at different settings. Equilibrate shoot for 5 minutes at each setting before making readings	Light intensity / Air humidity	Ambient light / Ambient humidity	Repeat experiment with all leaves removed

Respiration experiments

Aim	Independent variable	Dependent variable	Method summary	Controlled variables Variable	Controlled variables Value	Control
To determine the effect of temperature on the rate of locust respiration	Temperature locust maintained at, e.g. 15°C, 25°C, 35°C, 45°C, 55°C	Rate of respiration as shown by abdominal movements, e.g. number of abdominal movements/minute	Locust in respirometer allowed to equilibrate to temperature for 5 minutes Number of abdominal movements in 3 minutes counted to calculate mean/minute	Light intensity Volume available	Ambient Kept in same boiling tube for each reading	n/a
To determine the effect of temperature on the rate of respiration of yeast	Temperature at which yeast maintained, e.g. 20°C, 35°C, 50°C, 65°C, 80°C	Time taken for methylene blue to turn colourless/minutes	Yeast and sucrose solution equilibrated separately to temperature and mixed with methylene blue; maintained at given temperature and timed for disappearance of blue colour	Concentration of yeast suspension Volume of methylene blue	10g/100 cm³ suspension 0.1 cm³	Yeast suspension boiled for 10 minutes and cooled before using at each temperature

Field work experiments

Aim	Independent variable	Dependent variable	Method summary	Controlled variables Variable	Controlled variables Approx. value	Control
To determine the effect of light intensity on the % cover of ground ivy along a transect	Light intensity given by distance along transect from a wooded area into a field	% cover ground ivy	Set up 5m transect from a wooded area into a field At 1m intervals assess % cover of ground ivy in a 0.5 m × 0.5 m gridded quadrat	Air temperature Ground humidity	20°C 70%	n/a
To compare the effect of fast and slow water flow rate on the number of mayfly nymphs in a stream	Water flow rate – still water or approximately 2 m s⁻¹	Number of mayfly nymphs	Find 10 sites in still water and 10 sites on water flowing at approximately the same speed and place 0.5 m × 0.5 m open frame quadrat on stream bed Kick sample for 2 minutes and count mayfly nymphs retrieved in net after 30 seconds	Water temperature Water depth	16°C 0.1m	n/a

B.4 Writing up an experiment

When you have done an experiment or investigation in the lab or the field, you may be asked to formally write up part or all of it. When you write a practical report, you should consider the following headings:

1. Aim
2. Prediction
3. Experimental design
   - Details of the independent variable
   - Details of the dependent variable
   - Details of controlled variables
   - Repeatability
   - Control experiment
   - Safety risk assessment
4. Results
   - Suitable table
   - Graph
   - Statistical test, if appropriate
5. Analysis of results
   - Trend
   - Consistency of raw data
   - Sources of uncertainty plus improvement
   - Explain/relate results to theory
   - Conclusion
6. Plan for further work
7. References

If you write about each and then leave a line before moving on to the next, your thoughts will be organised and each page will look organised too. You could write each heading to help you focus your thoughts. For each paragraph, as you write, ask yourself, 'Is what I am writing now directly related to the subheading?' If the answer is 'No', think again.

Explanation of the headings

1. Aim

You must say what your experiment is designed to find out and identify the independent and dependent variables.

independent variable   *dependent variable*

Example

Aim: To investigate the effect of potato disc number on the volume of oxygen produced by catalase in hydrogen peroxide breakdown.

If you are writing about an enzyme experiment, it is important to name both the enzyme and the substrate in the aim, rather than just saying 'enzyme' or 'substrate'.

Do not write:

'Aim – To test the effect of enzyme concentration in the rate of digestion of substrate.'

You should phrase it as:

'Aim – To test the effect of trypsin concentration in the rate of digestion of milk protein.'

Make sure that your aim refers to what is actually plotted on the graph. Many people aim to test a rate but then plot time instead. It doesn't matter which you do but you must be consistent.

2. Prediction

A prediction links the two variables with the direction of the effect. You should say what you think might happen. If you are making a measurement, it is very hard to give a numerical prediction, but if you are testing an effect you can predict a trend. For example: 'Prediction – As the number of potato discs increases, the volume of oxygen produced when catalase digests hydrogen peroxide will increase.'

>> *Pointer*

Keep the aim simple by just testing one factor. Multiple aims make an experiment too complex.

3. Experimental design

A. The independent variable

Even though you have already stated your aim, you must say explicitly, 'The independent variable is...' and then you must give a minimum of five values for it. One of the values could be zero, e.g. for an enzyme concentration, but it is still a value so you must state it.

> **Example**
>
> **Aim:** to test the effect of lead nitrate concentration on the length of mung bean radicles four days after germination.
>
> **The independent variable** is the concentration of lead nitrate.
>
> Its values will be 0, 0.05, 0.10, 0.15 and 0.20 mol dm^{-3}.

5 values   *don't forget the units*

NB 1: if you are using hydrogen peroxide, state its concentration in the units 'vol'. It decomposes to give oxygen and the more hydrogen peroxide you have in solution, the bigger the volume of oxygen that is made:

1 dm^3 1 vol hydrogen peroxide gives 1 dm^3 oxygen

1 dm^3 2 vol hydrogen peroxide gives 2 dm^3 oxygen

NB 2: using the unit % when describing a solution has a problem because it can mean two different things. If you are diluting a stock solution to get different concentrations for your independent variable, make it clear that a 10% solution is 10% of a stock solution. The confusion arises because 10% could also mean 10 g per 100 cm^3 solution. Be very clear which you mean.

NB 3 Field work:

i) if you are comparing two habitats, obviously you can't give five values for the independent variable. Say what the two habitats are and give approximate values if relevant.

> **Aim:** to compare the leech numbers in still and flowing water.
>
> **The independent variable** is the water flow rate and is either zero (in still water) or approximately 1 m s^{-1} in the stream.

ii) the only time you are justified in using the word 'amount' is if you are considering the amount of trampling.

> **Aim:** to test the effect of the amount of trampling on the percentage cover of the grass Yorkshire fog.

B. The dependent variable

This is what you will be measuring, counting or calculating. It will already have appeared in your aim and this is what you will plot on the *y* axis of your graph and discuss in your analysis. You must say what units you will use for it.

● ● ●**Practical tip**● ● ●

The units of concentration of hydrogen peroxide are 'vol'.

● ● ●**Practical tip**● ● ●

If you use % as a concentration, make it clear what % means. Is it g/100 cm^3 or is it a dilution of a stock solution?

 Grade boost

Only use the word 'amount' when you are testing the effects of soil trampling.

>> **Pointer**

Remind yourself whether you are writing about time or rate. It doesn't matter which but you must be consistent in the aim, prediction, results, graph, analysis and conclusion.

Example

Aim: to investigate the effect of trypsin concentration on the time taken to digest the casein in milk.

The dependent variable is the time taken to digest the casein, in other words, the time taken for the milk to change from white to clear. It will be measured in seconds.

But if the aim referred to the rate of reaction, not the time, then the dependent variable would be rate and would be measured in s^{-1}, because you would calculate it using the formula rate = 1/time. In this case, the graph would have rate of reaction plotted on the y axis and the discussion would all be about rate, not about time. Do not discuss both and do not keep changing which you are writing about. You will lose a mark if you are not completely consistent.

In a field work experiment, you might be calculating a biotic index as your dependent variable. The commonest are Disney's, Simpson's and the Lincoln index. There are no units for these because they are arithmetical devices using counts, so you would need to explain that.

C. Controlled variables

The only factors that should change in an experiment are:

- The **independent variable**, for which you choose the values.
- The **dependent variable**, for which you measure, count or calculate the values.

Everything else must be kept constant so you can be sure that it is only the change in the independent variable that is affecting the dependent variable. The factors you choose to control deliberately are the **controlled variables**. You should describe the most important controlled variables and state the values you have chosen.

• • • **Practical tip** • • •

Many experiments have temperature as a controlled variable. You use the thermometer to measure the temperature, not to keep it constant.

Example 1

Aim: to test the effect of pH on the time taken for amylase to digest of starch.

Controlled variables:

The amylase concentration will be kept constant at 0.1%, i.e. 0.1 g per 100 cm^3 amylase solution.

The concentration of casein will be kept constant by using solutions containing 4% milk powder, i.e. 4 g milk powder per 100 cm^3 of solution.

>> **Pointer**

It is not enough just to say the pH will be kept constant. You must state what the pH is.

Do not say that the temperature will be kept constant by using room temperature. This is much too vague, and in any case, room temperature changes. Always state a particular temperature. If you are using human enzymes, 37°C is a suitable temperature, but plant and fungal enzymes may have completely different optimum temperatures and so 37°C may not be suitable for them.

In the example above, two important factors are given as the controlled variables. Of course there are others that could be used, such as the volume of amylase, the volume of milk and the temperature.

Example 2

Aim: to test the effect of urea concentration on the time taken for urease to digest urea.

The most suitable value is not always pH7. If you use pepsin, a suitable pH is 2.

Controlled variables:

1. pH will be kept constant at pH7 by using a pH7 buffer.
2. The temperature at which reaction is carried out will be 30°C, using an electronically controlled water bath.

If you are doing field work, you must explain that it is not possible to control the environment, so you would choose sites where the environmental factors are as similar as you can get them. You would then measure them to be sure that, while unlikely to be identical as they would be in a lab experiment, they are close enough to have no significant effect on the dependent variable.

Field work example

Aim: to test the effect of soil pH on the density of dog's mercury.

Light intensity and air temperature are significant factors in controlling plant growth so it would be sensible to have those as the controlled variables. You would set up a 10 m × 10 m sampling area and place 15 quadrats at random coordinates for counting dog's mercury. To monitor the dependent variables, at each quadrat you would measure light intensity and air temperature. The readings in each quadrat should be similar.

D. Repeatability

In your design, you need to state that you are repeating the readings for each value of your independent variable. The more readings you take, the more reliable the mean is. At this level, three readings are usually enough, but if one of them is very different from the other two, you may decide to take four. The very different reading can be discarded, as it is anomalous.

You need to explain why you are making the repeats and explain that it diminishes the uncertainty in your measurements and increases the confidence you have in drawing conclusions from the data. If all your readings are close, then your readings show repeatability. The spread around the mean is a measure of the repeatability and is another way of describing reliability. It will be shown on the graph by the range bars or, for some field work experiments, the standard deviation bars you draw.

In field work experiments it is not possible to make exact replicate readings. You can never step into the same river twice. You should explain this to your examiner, and explain that your reliability is enhanced by representative sampling, and the number of readings taken is accounted for in the statistical test.

E. Control experiment

The control experiment shows it is the change in the independent variable that causes the change in the dependent variable. The controlled variables are an attempt to show this, but it is the control which provides the verification. It often involves inactivating an essential component of the experiment, e.g. by denaturing an enzyme.

Example 1

Aim: to test the effect of pH on the time taken for pepsin to digest egg white.

Control experiment: the pepsin solution was boiled for 10 minutes to denature the enzyme and then it was cooled before use.

NB don't forget to write that you cool the enzyme before using it.

In Example 1, the active agent in the experiment, i.e. the enzyme, was removed and so no digestion is expected, showing that the pepsin is responsible for the digestion. You must also say that to show that it is the pepsin that is

Grade boost

Remember to comment that the more readings you take, the more reliable the mean is.

Pointer

If you are making counts in a quadrat, you can work out when you have taken enough readings. You recalculate the mean with every additional result. This is called a running mean and you will know when you have enough readings because the running mean will stop changing as you take more readings.

responsible, all other conditions must stay the same, i.e. egg white volume and concentration, pepsin volume, temperature, pH of solution. You could just say 'all other conditions remain the same' but specifying them produces a more thorough write up.

Example 2

Aim: to test the effect of light source distance on the volume of oxygen emitted by algae suspended in alginate balls.

Control experiment: the algal suspension was boiled for 10 minutes to kill the algae. They were cooled before being suspended in alginate balls.

Grade boost

Don't forget to mention that after you boil an enzyme or algae for the control experiment you have to cool them before use.

Example 2 shows that the algae are responsible for the changing volumes of oxygen evolved.

As explained on page 33 a control is not a zero value of the independent variable, e.g. distilled water instead of enzyme or no light in a photosynthesis experiment. The control must specifically remove the active ingredient.

In some experiments, there is no control because there is no active ingredient to remove. This is often the case when you are making a direct comparison:

Example 3

Aim: to determine the species diversity in long grass and short grass areas.

Example 3 has no suitable control because removing the grass has no meaning in this experiment. If this were your experiment, you must explain why no control is possible and then discuss what a control is used for.

F. Risk assessment

There are three things you have to write about in the context of the risk assessment. You must:

- identify the hazard
- say why it is a hazard
- say what you would do to minimise the risk.

Sometimes, students do not know the difference between a hazard and a risk.

- The hazard is the item that might cause you harm, e.g. a rusty scalpel blade could cause septicaemia if you cut yourself; or the ethanol next to a flame might cause a fire and burn you.
- The risk describes the activity that might cause the harm, e.g. using a scalpel in dissection; or, when performing a starch test, if a flame has not been extinguished before pouring ethanol to decolorise your leaf.

Pointer

If your experiment has no hazards and there are no risks, say so. That is still as risk assessment.

You must identify the most significant hazards in your experiment, so do not write about tying back your hair, wearing a lab coat, putting your bag under the bench or not running in the lab. This is altogether far too Year 7.

Here is a list of types of experiment and some of the hazards and risks they present. For your particular experiment, you must explain why they are hazards and you must explain the risks, i.e. the part of your experiment in which they may cause harm. Suggestions for how to minimise harm are in the fourth column. Your experiment, however, may be about something else entirely and you will have to consider its hazards and risks.

Experiment type	Major hazard	Risk from major hazard	Hazard control
Enzymes	Allergic reaction of skin in response to foreign protein	Dispensing enzyme solution	Avoid skin contact and wash with soap and water if contact occurs
	Proximity of water and electricity could result in electric shock	Using electronic water bath	Place water bath at distance from electric socket; take care not to splash water; immediately mop up any spilt liquid
Catalase	Cutting potato discs with sharp scalpel could result in cuts to skin	Preparing potato discs	Care with dissecting equipment; if skin cut or pierced, wash with soap and water immediately
	Hydrogen peroxide is an irritant to eyes and skin	Filling syringe barrel and placing it on apparatus	Avoid skin contact and wear goggles
Photosynthesis	Skin may touch hot lamp	Moving lamp to adjust light intensity	Avoid skin contact with bulb; 10 minutes in cold water if burned
	Eye damage from halogen light source	Looking at light source	Avoid looking directly into lamp
Bacterial cultures	Infection with pathogens	During culturing and assessment of growth	Use sterile technique; do not open dishes once set up; incubate culture below 30°C; autoclave dishes prior to disposal to kill bacteria
Field work at coast	Slipping on rocks may cause cuts, strain or sprain	Working on loose or algal-covered rocks	Wear suitable shoes; walk very carefully; work within earshot of others in case of accident
	Getting cut off by tide coming in	Working in caves or coves	Check tide tables before beginning work; work within earshot of others
Field work using bodies of water	Drowning by water inhalation	Accessing stream or lake	Walk very carefully; work within earshot of others in case of accident
	Weil's disease from ingesting *Leptospirosis* in water	Taking readings in stream or lake	Use rubber gloves to avoid contact with water; keep cuts covered
Terrestrial field work	Tripping on uneven terrain or plant roots	Accessing sample sites	Walk very carefully; work within earshot of others in case of accident
	Poisonous plants and animals may sting or breach skin	Collecting data	Avoid ingesting any biological material; wear appropriate clothing to keep skin covered
	Getting lost in unfamiliar area	Investigating new site	Work within earshot of others; carry mobile phone with emergency contact numbers

>> *Pointer*
To protect your eyes, you wear goggles, not googles.

>> *Pointer*
'Hazard' has one z. With two it is an American TV programme from the early 1980s.

>> Pointer

Write units in the column headings only. Do not write them in the cells of the table. The cells have numbers only.

⟰ Grade boost

The convention these days is not to put the units in brackets, but to write an oblique after the heading and before the unit, as the example below shows.

4. Results

The word 'results' is used for both raw data, which are the readings you take, and for processed data, which are the calculations you make from the readings. Unless it is very clear which you mean, it is best to refer to 'readings' or 'calculated values' and avoid the word 'results'.

A. The table

i. **Column headings:** The way you head the columns is crucial. The column headings must be complete and units must be included in the column headings:

> **Examples**
>
> 'time for methylene blue to turn colourless / s', not just 'time / s' or 'volume of oxygen produced in 5 minutes / cm³' , not just 'volume / cm³ '

ii. **Arranging the table:** You will need a column for the independent variable and then four columns for the dependent variable. These four are three for the three readings and one for the mean:

Example

Aim: to test the effect of temperature on colony diameter of cultures of *Streptococcus albus*.

Results:

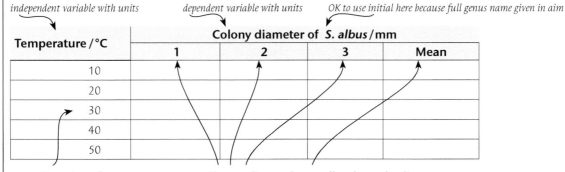

| Temperature / °C | Colony diameter of *S. albus* / mm | | | |
	1	2	3	Mean
10				
20				
30				
40				
50				

independent variable with units — *dependent variable with units* — *OK to use initial here because full genus name given in aim*

Five values of independent variable

Three readings and mean all under one heading, which is the dependent variable with its units

iii. **What type of average?:** 'Mean' is a better word than 'average'. If you write the word 'average', it will be assumed that you mean mean, but in fact, you might mean mode or median. So choose words with care. If it is was a mean that you calculated, write mean. Remember to include the mean column under the heading for the dependent variable, so its units are clearly shown.

>> Pointer

Some people prefer to put the mean in a different table because it is a calculated value rather than a reading. If you do, remember to head the independent variable column correctly and to include the heading for the dependent variable.

iv. **Which units?:** Take care when you write the units that they make sense. If you are giving a length, mm or metres are best as they are SI units. For volumes, cm³ or dm³ are the commonest and, being SI units, are preferable to ml and litres. If you have calculated an area, make sure the units are correct. It is too easy to write mm³ when you mean mm².

v. **Units and decimal places:** The most important thing is BE CONSISTENT. But also be sensible.

The thermometer may have calibrations one degree apart. All you write then is the number, e.g. 20°C, not 20.0°C. If you go to one decimal place, the implication is that you can read to ±0.1°.

You may use an electronic timer that gives the time to ±0.01s. Sadly your reaction time is not this good so it is meaningless to give times to this degree of accuracy. Depending on the experiment, the nearest second may be suitable, e.g. if you are timing how long it takes to collect 10 cm³ oxygen. But if you are timing how long it takes for methylene blue to decolorise, it is impossible to be that precise because the colour change is gradual. In this case, it may be better to read to the nearest minute. As the SI units of time are seconds, if you choose a different unit, such as minutes, explain underneath your table what you have done and why. It will not give you more marks, but it shows your examiner that you understand what you are doing and it makes a better report.

> **Something to avoid**
>
> If you are recording time use one unit only. It may be seconds, it may be minutes but what it must not be is minutes and seconds. Sometimes people write 1.30 minutes. This probably means 1 minute 30 seconds but it may mean 1.30 minutes. Then when a mean is calculated, even if it meant 1 minute 30 seconds people treat it as 1.30 min and then their mean is wrongly calculated. All too confusing. Stick to one unit only. Do not mix.

Sometimes there are no units

If your experiment uses readings from a colorimeter, there are no units because the reading is a comparison. Then you must write AU or arbitrary units, and explain why.

If you are plotting a count then there will be no units. You would then label the axis 'number of ....' so it is clear that there are no units.

vi. **The mean and decimal places:** When you calculate the mean, in theory you should use the same number of decimal places as in your readings, because a calculated value cannot be more accurate than the readings that produce it. But it is acceptable to use one more decimal place if it adds clarity. Once again, underneath the table, explain what you have done and why.

Except for the mean in some cases, the number of decimal places must be the same throughout the results table. Whether or not the mean has a different number of decimal places from the readings, make sure that within each vertical column, you always have a constant number of decimal places.

vii. **Repeats:** Three is the best number. If for any reason, you have not done three, explain why underneath your table. In some cases, your teacher may provide you with another set of results to add to your own. Make sure it is very clear which are yours and which have been given to you.

● ● ● **Practical tip** ● ● ●

If you are recording time, stick to one unit only. Do not mix minutes and seconds.

B. Graph

i. **Axes:** Label both axes correctly and fully, just as in the column headings of the results table. The independent variable goes on the horizontal axis and the dependent variable goes on the vertical axis. If you are plotting means on the y axis, the label must say mean. It is truly alarming how many people forget to write 'mean' here.

> **Example**
>
> **Aim:** to test the effect of alcohol concentration on the heart rate of *Daphnia*.
>
> With three readings at each alcohol concentration, you would calculate a mean heart rate to plot. The vertical axis would be labelled 'mean heart rate of *Daphnia* /beats per minute'.

You don't need to write x or y on the axes and you don't need to put an arrow at the end of each.

ii. **Scale:** Choose the scale carefully so that the data points that you plot will produce a line that covers at least half the available space on both axes.

Like this: **Not like this:**

Make sure you have put the correct units on the axes. Check they are the same as in your table.

Choose scales that are easy to read. Choose scale divisions that divide by 5 or 10, not by three.

Like this:

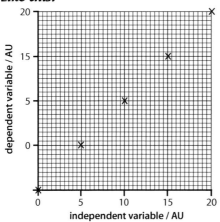

Not like this:

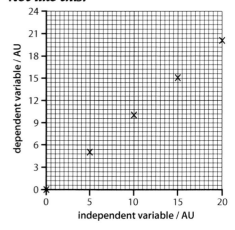

iii. **Origin:** whether or not the scale starts at 0, there must be a number at the origin on both the x and the y axis.

iv. **Break lines:** can be used to make sure your data points cover a large part of the graph paper:

But you could avoid them by having the x axis start at 50°C:

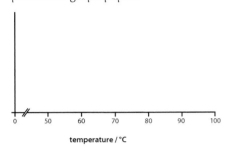

temperature / °C

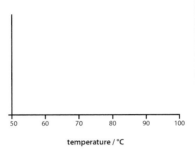

temperature / °C

Pointer

Do not start your scale, then use break lines and then continue with the scale. The break line comes immediately after a 0, or not at all.

v. **Points:** All the points must be plotted accurately. This is easier if you have chosen a sensible scale.

vi. **Lines:** Join the data points individually with a ruler, making sure you go from the centre of one cross to the centre of the next. Make sure you use a sharp HB pencil and that you rule it properly the first time and don't need to erase the line to try again.

In Biology, it is the data points that are significant, so make sure the line does not obliterate them. The line is just a construct of your imagination. Unlike in Physics or Chemistry, we cannot assume a straight line relationship between the variables so a line of best fit is only occasionally justified. Because you only know what has happened in the range at which you measured, you are not justified in taking the line beyond the data points. So no extrapolation. This means that you cannot draw the line beyond the points at either end. Do not join the line to the origin unless there is a data point there.

Pointer

When you plot the points, make them visible. Dots disappear when you draw a line. Crosses are best with the centre of the cross at the exact coordinate.

Grade boost

For the graph line, join point-to-point with a ruler. No extrapolation.

Like this: **Not like this:**

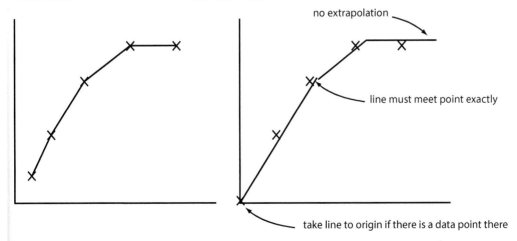

no extrapolation

line must meet point exactly

take line to origin if there is a data point there

vii. **Experiment with line of best fit:** If you are reading an intercept from the graph, a line of best fit is suitable. Make sure you explain why you have chosen a line of best fit, rather than joining the points.

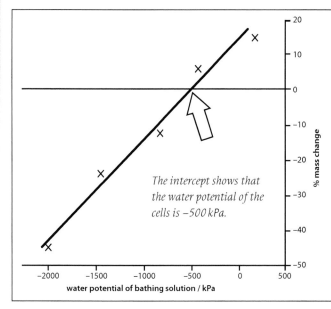

The intercept shows that the water potential of the cells is −500 kPa.

water potential of bathing solution / kPa

Example

Aim: to measure the water potential of potato cells.

You could measure the mass of potato chips before and after incubation in solutions of different water potentials. If you plot % mass change against water potential, the line will cross the x axis at the water potential of the cells. You can draw a line of best fit for this because the line will take into account the inaccuracy of each point so your intercept can be read more accurately than if you had taken only the two points either side.

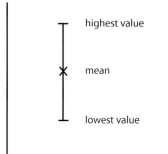

highest value

mean

lowest value

viii. **Use of error bars:** The commonest type of error bar at A Level is the range bar. You use this when each point on your graph is a mean value. You plot the highest and lowest readings for each, in addition to the mean. When you do this, draw a cross for your mean and two little bars for the highest and lowest readings.

Sometimes, people use crosses for all three and then the graph is much more difficult to interpret, so don't do that.

If you have done a field work experiment, it may be suitable to plot a bar chart showing two mean values and you can use you can use range bars here too:

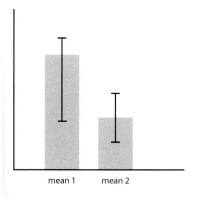

mean 1   mean 2

If you have done the sort of experiment that lets you plot a bar chart, you are likely to have done a t-test, in which case you will have calculated standard deviation. Instead of plotting range bars, you could plot standard deviation bars instead.

It doesn't matter which you do but remember to write somewhere on your graph 'bars show range' or 'bars show standard deviation'.

If you have done a field work experiment where you are testing a correlation, you would plot a scatter graph and so there will not be any error bars:

Example

Aim: to test the effect of soil pH on the percentage area cover of heather.

The scatter graph might look like this:

Although soil pH is the independent variable, you can only take results where you find them and as you are unlikely to find two patches of soil with exactly the same pH, there are no precise replicates.

So you cannot draw range bars.

It isn't appropriate to draw a line of best fit through these points because you cannot assume a straight line relationship. A statistical test will tell you if the assumption is justified.

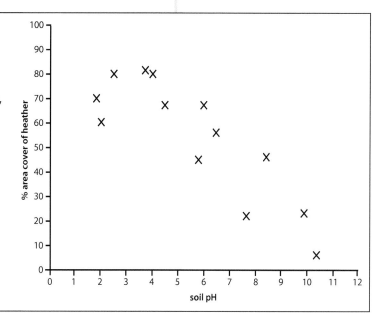

C. Statistical analysis

If you do a fieldwork experiment, you may analyse your data with a statistical test. Cynics say that you use statistics only when the answer isn't obvious. But that isn't true. You use statistics to analyse field data because genetic variation between organisms and the minute differences in abiotic factors prevent you from making true replicate readings. So without statistics, it would be hard to draw a valid conclusion.

When you present your statistical analysis, there are six things you have to do:

1. Explain why you have chosen this particular test – see pages 42–43.
2. State the null hypothesis.
3. Calculate the test statistic correctly.
4. Say how many degrees of freedom you are working to and explain that you are using a probability of 0.05 or that you are using 95% confidence limits.
5. Quote the critical value of the test statistic from the tables and say whether your calculated value is higher or lower.
6. Say whether you accept or reject the null hypothesis at the 0.05 level of significance.

5. Analysis

A. Trend

Describe what your graph tells you. First of all describe the general trend and say if it appears to be a direct relationship, i.e. straight line with a positive gradient, and say if it seems to go through the origin (Example 1). Or say if it is an inverse relationship, i.e. a straight line with negative gradient (Example 2). If the gradient seems to change, describe that (Example 3).

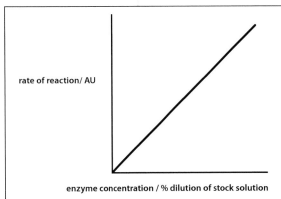

Example 1

Aim: to investigate the effect of enzyme concentration on the rate of reaction of amylase.

Trend: as the enzyme concentration increases, the rate of reaction increases. There is a positive gradient and the line goes through the origin.

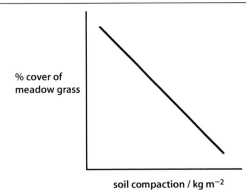

Example 2

Aim: to test the effect of soil compaction on the percentage cover of meadow grass.

Trend: as the soil compaction increases the percentage cover of meadow grass decreases. There appears to be an inverse relationship between the two.

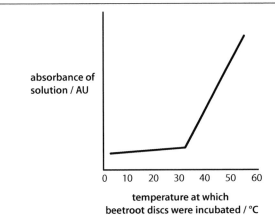

Example 3

Aim: to test the effect of temperature on the permeability of beetroot membranes.

Trend: there is a gradual increase in membrane permeability as the temperature increases up to 30°C. Above 30°C the gradient of the line increase sharply showing that permeability increases greatly so 30°C can be recognised as the transition temperature for this membrane.

For a field work experiment, you could state the conclusion derived from your statistical test:

> **Null hypothesis:** there is no difference in the mean length of the water shrimp body in deep and shallow bodies of water.
>
> **Trend:** the calculated value of t is greater than the critical value of t at 38 degrees of freedom and 0.05 probability, so the null hypothesis is rejected at the 0.05 level of significance.

The trend you describe must refer to your own results and not what you expected to happen. Refer to your graph or table while describing the trend so that you describe what you actually found, rather than what the textbook says.

Word of warning: if your graph has a peak, only describe it as an optimum if you mean it is the best:

 Grade boost

If you have unexpected results, describe them accurately and try to explain why you got them.

An experiment to test the effect of temperature on enzyme activity might produce a graph with a peak. If that peak occurred at 45°C, you could say that the enzyme has an optimum temperature of 45°C.

But an experiment testing the effect of temperature on the number of aphids living on a bean plant might give you the same shape graph. You would not describe the temperature giving the highest number the optimum, because aphids destroy bean plants so there is nothing 'optimum' about the temperature that gives the highest number.

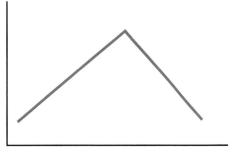
temperature / °C

B. Consistency

Here you look at your raw data and see how similar the replicates at each value of the independent variable are. This is also described as reliability or repeatability:

> **Example**
>
> **Aim:** to assess the effect of water potential of a bathing solution on the percentage of cells that are plasmolysed.
>
> In this experiment, you would float rhubarb epidermis in sucrose solutions with different water potentials for 30 minutes and then mount them on a microscope slide in a drop of that solution. Using your microscope, you could look at 100 cells and count how many of them are plasmolysed. You would make three separate slides at each water potential and these are the replicates.
>
> | Water potential of sucrose solution /kPa | % cells plasmolysed | | | |
> |---|---|---|---|---|
> | | 1 | 2 | 3 | Mean |
> | 0 | 2 | 3 | 0 | 2.5 |
> | −500 | 34 | 40 | 38 | 37.3 |
> | −1000 | 58 | 60 | 59 | 59.0 |
> | −1500 | 70 | 78 | 64 | 70.1 |
> | −2000 | 97 | 89 | 93 | 93.0 |
>
> **Results:** Look at the replicates in the results table and decide which water potential has the most similar and the most different results. Then state it like this:
>
> The readings at −1000 kPa are the most consistent because they are all close together. The mean at −1000 kPa is therefore the most reliable.
>
> The readings at −1500 kPa are the least consistent as they have the greatest difference between so the mean at −1500 kPa is the least reliable.

C. Use of error bars

If you have done a field work experiment where you are testing a correlation, you won't have precise replicates and so you can't draw error bars. You will need to explain why you do not have them and make a statement about how they can be used.

If you have an experiment with replicates, you can plot range bars or, with enough data, you can plot standard deviation or standard error bars. Just make sure you say which you have used. There are two ways you can use the bars and these relate to their length and to their overlap.

Range bar length

This is another way to think about consistency. The range bars show the highest and lowest readings and the mean falls between them. If the bar is short, then there is not much difference between the lowest and highest readings and so the readings are repeatable or consistent and so the mean is reliable. If the range bar is long, there is a big difference between the highest and lowest readings and so the readings are not repeatable or consistent and so the mean is not reliable. Look at your graph and decide which are the shortest and the longest range bars and use these to determine which are the most reliable and unreliable means.

Range bar overlap

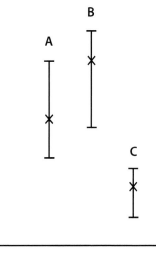

This is all about how truly different the means are. The graph shown here has three data points. These are means and the range bars are plotted around them:

The range bars for A and B overlap. This is because the highest reading for A is higher than the lowest for B. This means that we do not have confidence that the means for A and B are truly different.

The range bars for A and C do not overlap because the lowest reading for A is higher than the highest for C. Therefore we have confidence that the mean for A is truly higher than the mean for C. The same is true for B and C: the mean for B is truly higher than the mean for C.

A word of warning: the raw data are fundamental. The raw data determine the length of the range bars. So when you write about this you must explain the logic the right way round. You can say 'the raw data are consistent therefore the range bars are small' NOT 'the range bars are small therefore the raw data are consistent'.

Grade boost

When you are describing consistency there are two things you can talk about:

- Similarity in raw data.
- Length of range bars.

D. Uncertainty and improvement

This section is all about the inherent problems with your equipment or the design of your experiment. It is not about human error – the assumption is you know how to use all the apparatus and you make precise readings with no errors. You are perfect but the experiment is not.

When you have identified the areas that lead to lowered confidence in your results or decreased accuracy in your readings, you can suggest suitable improvements.

Subjective or objective?

The biggest source of error is subjectivity. This problem was considered in the context of experiment design on pages 33–34. A reading is subjective when you are making a value judgement such as timing exactly when the colour of methylene blue has gone or just how red the hydrogen carbonate indicator is. When making a subjective decision about timing, a more accurate stop clock won't help. You may be able to measure to 0.001 s instead of 1 s but as you are making the choice yourself when to stop the stop clock, the increased accuracy of the clock won't tell you more accurately when the colour has gone. A more useful improvement when judging a colour change is to use a colorimeter. You could, for example, time how long it takes for your solution to reach a pre-determined absorbance. Then, as you are reading numbers on a scale, there is no personal choice involved and so the reading is objective.

Counting

A similar source of error involves counting. It is possible to count incorrectly. A field work example is testing the effect of phosphate ion concentration on the number of water shrimp. If you have a lot of water shrimp in your river sample and they are swimming fast, it is very hard to count them. In this case an improvement is to take a representative volume of your sample, e.g. 100 cm³ river water, and count how many water shrimp are there. Then measure the volume of your total sample and multiply up.

Accuracy of equipment

Your experiment may be entirely objective. In that case you must consider the accuracy of your equipment. Consider all the measurements you have made. They may include temperatures, volumes, lengths or masses. The source of uncertainty is derived from the accuracy of calibrations. This was discussed a bit in the section on decimal places in the results tables (page 55) in relation to reading the scale on a thermometer. If the thermometer is calibrated in degrees, it is only accurate to 1C°. An improvement would be to have a thermometer which is calibrated to 0.1C°.

You can think about rulers and measuring cylinders in a similar way. A ruler may be calibrated in mm so to measure a length more accurately you could use callipers, calibrated to 0.1 mm. But you may be measuring the diameter of bacterial colonies, in which case, callipers would not be suitable. Here, you could project an image of a known length on to a monitor and calculate its magnification. Then you project an image of the colony and measure the diameter of that. Use the calculated magnification of your known length to calculate the actual diameter of the colony. It would still be to the nearest mm but the larger the length you measure, the smaller the percentage error.

>> **Pointer**
Objective is the opposite of subjective. An objective reading will be the same whoever makes it, whereas a subjective reading may depend on the person making it. Scientific method relies on people being able to repeat experiments, so objective measurements are essential.

>> **Pointer**
Make sure you know the difference between subjective and objective readings. Make sure you know which yours are.

>> **Pointer**
A thermometer calibrated in degrees is accurate to +1C°. You read it to the nearest degree so your reading is accurate to ±0.5C°.

>> **Pointer**
You should only write °C if it is an actual temperature. If you are talking about a number of degrees, it's the other way round: C°. For example, if the temperature increases from 20°C to 30°C, it has increased by 10C°.

>> Pointer

Be realistic when deciding how many decimal places to use. You are not working on the Large Hadron Collider. Going to an accuracy of three decimal places in your equipment may not make the experiment any more accurate because of all the other sources of error.

A measuring cylinder may be calibrated in cm^3 so an improvement would be to use a burette, calibrated to 0.2 cm^3 or a syringe calibrated to 0.1 cm^3.

When measuring mass, your balance may be correct to the nearest 0.1 g so a balance that measures to 0.01 g would make the reading more accurate. Another way of improving mass measurements applies if you are using a top-pan balance. If someone walks by while you are using the balance, air movements may lift the balance pan. It may be too small an effect for you to see, but the effect will be to give you an artificially low mass. So an improvement would be to shield the balance from air movements.

Sources of error are obvious once you know how to look for them. In all these examples discussed above, the improvements could be worked out from a careful examination of your equipment. Your examiner is interested in your ideas about inaccuracy, rather than accuracy. So decide on the biggest source of error and describe it in detail. Give two examples of improvements to your experiment, one related to this major source of uncertainty and one other.

Photosynthesis example

A popular experiment involves lining up vials of photosynthesising algae at various distances from a lamp. Major sources of error here are shading and, in the case of incandescent bulbs, differential heating effect from the lamp. The best solution to the shading problem is to do each vial separately. This takes much longer but this discussion is theoretical only, so that would be a valid improvement. If you have a fluorescent lamp, there should be no heat coming from the bulb, so that would be an improvement. Another way around this is to put a container of water between the lamp and the vials. This would absorb the heat and not the light. But take care. If you use a beaker of water, the lensing effect of the curved surface will alter the light intensity in different ways at different distances. If you are going to do this, you need a glass vessel with flat sides, like a chromatography tank.

Two field work examples

In a field work experiment using a t-test, you could explain that the use in Biology of the 0.05 probability contributes to your acceptance or rejection of the null hypothesis and so enhances the accuracy of your conclusion, compared with if you had used, say a 0.1 probability.

A suitable improvement for an experiment using a quadrat would be to ensure you have the right area quadrat. You would have to do this before you start the main experiment, and it is a small experiment in itself, as described on page 40.

What not to do:

- Do not suggest more replicates as an improvement. It is true that the more replicates you have, the more reliable the mean but this is about experimental technique and more replicates do not make your results or your conclusion any more accurate.
- Do not suggest more values for the independent variable unless you are trying to identify a special property of the system. Here are some special cases where more values of the independent variable are justified:

First example: If you are looking for an optimum value, e.g. when testing the effect of temperature on enzyme activity, your graph may look like the one on the right.

The optimum temperature for this reaction lies between 30°C and 50°C so an improvement would be to take more readings at temperatures between these two to more accurately find the optimum , e.g. 35°C, 38°C, 42°C and 45°C.

It would not be an improvement to take more readings below 30°C or above 50°C as that would not provide any useful information.

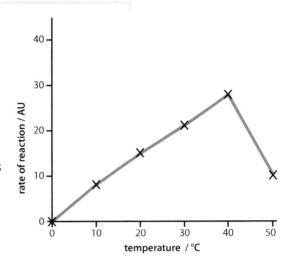

Second example: Some experiments produce a graph with a sudden gradient change; for example, looking at the permeability of beetroot membranes by assessing the absorption of pigment solutions leaked out after treatment in different temperatures. The graph may look like the one shown here.

The sharp gradient change represents a phase change in the membrane and you could improve the experiment by taking more readings around the temperature at which the gradient changes, to find that transition temperature more accurately. It will be between 20°C and 40°C so you could try 24°C, 28°C, 33°C and 37°C.

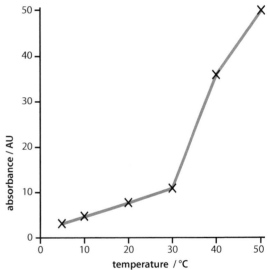

Third example: If the gradient were constant throughout, there would be no point in taking readings at more values of the independent variable within the range you have already covered. The only reason to do more values of the independent variable would be to extend the range if you had a good reason to think the gradient might change. If you were investigating the effect of substrate concentration on reaction rate, you might expect the rate to become constant when the enzyme is saturated. But the graph may not have reached a plateau at the substrate concentrations you have used, as shown here.

As an improvement it would be reasonable to try higher concentrations to find the substrate concentration that saturates the enzyme. So you might try 1.5 mol dm^{-3}, 1.75 mol dm^{-3} and 2.0 mol dm^{-3}.

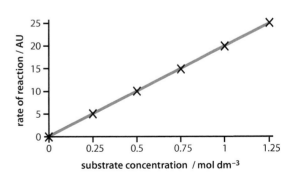

Grade boost

Remember to include an equation and description of the reaction in an enzyme experiment.

Grade boost

When you describe the theory that explains your experiment, relate it to your own results. Use phrases like, 'as can be seen from the graph...' or 'the table shows...' so that it is clear that you are interpreting your own data.

E. Explanation and relationship of results to theory

This is the biggest part of the analysis and it is where you show that you understand what you have done and its significance. Relate your own results to your theoretical expectations. Say if your results support the theory. Your results do not have to conform to theoretical expectations but you must be able to discuss them and try to explain them in the context of the major biological principles that underpin the experiment.

Some examples:

- If it is an experiment looking at the decolorisation of methylene blue by yeast, you must explain the biochemistry of both aerobic and anaerobic respiration. Both will be taking place unless you have made your solutions in boiled water and stored them under oil and performed the experiment in a nitrogen atmosphere.

- If your experiment tests the effect of light on photosynthesis, then you will need details of the light-dependent reaction of photosynthesis. You will need to explain the concept of limiting factors and how that applies in this case.

- An experiment testing the effect of antibiotics on bacteria will need a description of bacterial cell walls and the effects of antibiotics on them.

- A field work experiment will need to describe the niche of the organism under test and explain how its physiology or anatomy or behaviour is relevant to the independent variable you are testing. If you have tested the effect of light intensity or monitored it as a controlled variable, it is important to realise that it is not the light intensity here and now that matters. What is significant is the average light intensity over the whole of the growing season. All you can do is measure it on the day you do your experiment, but you have to use that as a proxy for the whole of the time the plant has been above ground.

- If you have done an enzyme experiment, make sure you describe the reaction by giving an equation and explaining why it is physiologically important.

Use of language is very important here. Your choice of words must be such that your meaning is absolutely clear. You will find more ideas about clear writing on pages 74–75. Plan what you are going to write for this section and then divide it into paragraphs, where each paragraph contains one thought. Not only will it look more attractive, but it will show that you think in an organised fashion.

F. Conclusion

This wraps up the whole experiment and brings together various strands. At its simplest, the conclusion is a statement of what you have learned, i.e. the trend you have observed and is an answer to the aim. Here are three examples:

Aim	Trend
To test the effect of temperature on the trypsin digestion of milk.	As the temperature increases, the time taken for trypsin to digest milk decreases up to 40°C. Above 40°C, the time increases.
To test the effect of carbon dioxide concentration on the rate of algal photosynthesis.	The rate of photosynthesis increases as the carbon dioxide concentration increases up to 0.05 mol dm^{-3}. Above this concentration, the rate decreases.
To test the effect of dissolved oxygen concentration on the number of mayfly nymphs.	As the concentration of dissolved oxygen increases, the number of mayfly nymphs increases.

But you have already described the trend at the beginning of your analysis, so the actual concluding paragraph needs more. The conclusion needs to take this trend and comment on it in relation to various other aspects that you have written about, e.g.

a) Refer back to the prediction you made at the start and say if it is supported.

b) Illustrate the trend by making a percentage calculation if you observed a particular increase or decrease.

c) Refer to the data.

d) Refer to the graph.

e) Refer to the explanation you have given.

f) Comment on the reliability.

g) Comment on the accuracy.

>> **Pointer**
The conclusion is a distillation of the main points of your analysis. Check you have at least three of the items mentioned in the list given here.

6. Plan for further work

You must imagine using the same general experimental techniques as the main experiment, so there is no need to write the method. There are three things to do:

A. Choose a new independent variable and make the statement 'the new independent variable is …'.

>> **Pointer**
Further work allows you to find out more about the system you have investigated in your main experiment. Keep the general technique the same but use a new independent variable.

> **Example**
>
> **Main aim:** to test the effect of lactase concentration on the rate of digestion of lactose.
>
> **Further work independent variable:** the new independent variable is the temperature.

Avoid an unsuitable choice, such as investigating the effect of the volume of buffer, so that you obtain meaningful information.

B. Choose two controlled variables and give their values. One could be the same as in the previous experiment but the other must be different. It is logical to make the second controlled variable the optimum value of the independent variable of your main experiment.

> **Example (as above)**
>
> **Main aim:** to test the effect of lactase concentration on the rate of digestion of lactose.
>
> **Further work independent variable:** the new independent variable is the temperature.
>
> **Controlled variables:**
>
> Lactase concentration at 20% of stock solution (found from main experiment to be suitable).
>
> Lactose concentration, using powdered milk at 4 g / 100 cm³ solution (as in previous experiment).

C. Expected result – three things to think about here:
  a) A sentence to say what you think will happen.
  b) A sketch graph to illustrate your prediction.
  c) A brief explanation.

> **Example (as above)**
>
> **Main aim:** to test the effect of lactase concentration on the rate of digestion of lactose.
>
> **Further work independent variable:** the new independent variable is the temperature.
>
> **Controlled variables:**
>
> Lactase concentration at 20% of stock solution (found from main experiment to be suitable).
>
> Lactose concentration, using powdered milk at 4 g/100 cm³ solution (as in previous experiment).
>
> **Expected results:**
>
> a) As the temperature increases up to 30°C the reaction rate will increase. Above 30°C the reaction rate will decrease.
>
> b) Sketch graph:
>
>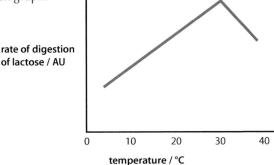
>
> c) Increasing temperature increases the molecules' kinetic energy so more successful lactase-lactose collisions increase the rate of reaction. Above 30°C, the additional energy produces intramolecular vibrations which break hydrogen bonds maintaining the active site, the enzyme is denatured and so the reaction rate decreases.

Grade boost

1. Don't forget to label the axes, including the units.

2. Only a brief explanation is expected – a few sentences are enough to show that you understand the principles involved.

7. References

It is important to read around any topic you are studying. When you do an experiment, it is useful to read the relevant theory. If you are measuring a parameter, e.g. the rate of water uptake into a leafy shoot at different light intensities, it is useful to find readings that other experimenters have made. You may then compare your readings with theirs. Or you may wish to discuss your results using information that is beyond the scope of your specification but is, none-the-less, relevant.

It is essential to avoid plagiarism so you should provide references for what you have read. They must be presented in such a way that someone else may easily retrieve them. You should include the author, date of publication, title, publisher and a page reference.

Supposing, for example, following an experiment on food chains in a fresh water habitat, you have used an old textbook to find out more details of the digestive system of the crayfish. You might cite your reference as:

Chapman, G. and Barker, W.B. (1964) *Zoology*, Longman, p. 155.

Or perhaps you wish to reference a journal article on the use of pressed plants in a herbarium for identifying plants in an investigation of species distribution in a habitat. Then you must give the author, date, title of the article, journal name, volume and number of the journal and page numbers:

Clark, R. (2009) 'The work of a herbarium curator', *Biological Sciences Review*, vol. 21, no. 4, pp. 37–40.

Remember that older books and journals have huge amounts of information and just because a reference is very recent does not necessarily mean it is very good. Books and journals written specifically for A Level students provide material at an appropriate level and in appropriate detail. It has also been reviewed and so you can be sure that it is reliable. This is often not the case in answers to questions that students submit online.

Pointer

Never ever copy word-for-word from a reference: your examiner will know. That is plagiarism and it is cheating. Always rephrase the information in your own words.

Writing up checklist

Heading	Design area	Detail	I can do this
Aim	Aim	Clearly state aim in terms of two variables	
Prediction	Prediction	Describe what you think will happen stating the direction of change	
		Do not include any explanation	
Design	Independent variable	Clear statement to identify	
		Five values	
	Dependent variable	Clear statement to identify what is plotted	
		Units	
	Controlled variables	Decide the most important and name them	
		Values for both, with units	
	Repeats	State how many; state they are used to calculate mean for reliability	
	Control experiment	Identify which feature is inactivated	
		Identify what stays the same	
		If no suitable control explain why not and what control is for	
	Hazard	Identify hazard	
		Describe the risk	
		Describe how to minimise risk	
		Describe what to do if hazard occurs	
Results	Recording	Detailed column headings	
		Units in column headings only	
		Enough repeats or an explanation as to why not	
		Appropriate data recording to suitable number of decimal places	
	Graph	Both axes correctly labelled in detail	
		At least half page used for data points and line	
		Correct units on both axes	
		Suitable linear scale with a figure at the origin for both axes	
		Suitable break lines if necessary	
		All points plotted accurately and easily seen	
		Clear line joining points with no extrapolation	
		Error bars if appropriate	
	Statistics in some field work or genetics experiments	Explanation of why this test has been chosen	
		Null hypothesis	
		Correct calculation of test statistic	
		Number of degrees of freedom	
		0.05 level of significance or 95% confidence limits quoted	
		State critical value from tables	
		State whether calculated value is higher or lower than critical value	
		State whether null hypothesis accepted or rejected at 0.05 level of significance	

Analysis	Trend	Describe trend or explain the meaning of the statistical test	
	Consistency	How repeatable the raw data are	
	Error bars	Use of the error bars or standard deviation	
	Uncertainty	Major source of error in equipment or procedure explained or, for stats, explanation of level of significance	
	Improvements	Two improvements to the procedure explained	
	Explanation	Explain results coherently including major biological principles	
	Conclusion	Combine knowledge, prediction, results, accuracy, repeatability to make a valid conclusion	
Further work	Independent variable	State explicitly the new independent variable	
	Controlled variables	State at least two controlled variables and give the value for each	
	Expected results	State expected results, include a sketch graph and give a brief explanation	

C: Supporting your work

C.1 Publications

You may have heard of the advice given to students in the United States when they have a problem with their work. Apply the four Bs in this order:

1. Brains (the most complex object in the known universe and we all have one).
2. Books (although now that includes all print and electronic media).
3. Buddies (our friends may know the answer but these days we are very conscious of the moral problems of copying and plagiarism, so take care how you use their work and ideas).
4. Boss (your teacher).

This section explores 2, i.e. the books, journals, websites and videos that you have access to. If you go straight to 4 and ask for an answer, the answer is all you have. If you look it up for yourself, you have the answer but you also have the intellectual development that got you there as well as all the snippets of information you picked up on the way. This is the beginnings of the differences between information, knowledge and wisdom. We would all like wisdom but you have to work for it.

There are many ways to acquire information and people learn best when they have identified their own preference. Using different types of learning for different types of knowledge is useful and the more ways you approach something, the more chance you have of finding the way that appeals to you. So don't use just one book or website or buddy. Try several and generate the answer that makes most sense to your way of thinking.

Pointer

Keep up to date with biology in the news. Read the science coverage in quality newspapers, e.g. the *Observer*, the *Guardian* or go to the Science section of the BBC News website.

Books

Textbooks for A Level students are written with a suitable amount of detail and with explanations at an appropriate level. This is why it is often much easier to find what you need in a textbook than on the internet.

The more times you read about a concept, the easier it is to remember. But you should also consult at least two textbooks, because the more varied ways you see a concept explained, the easier it is to understand.

Illuminate Publishing has produced textbooks and study/revision guides designed and written specifically for Advanced Level WJEC and Eduqas Biology:

WJEC Biology for AS Level
WJEC Biology for A2
WJEC Biology for AS: Study & Revision Guide
WJEC Biology for A2: Study & Revision Guide

Eduqas Biology for A Level Year 1 & AS
Eduqas Biology for A Level Year 2

Eduqas Biology for A Level Year 1 & AS Study & Revision Guide
Eduqas Biology for A Level Year 2 Study & Revision Guide
These are very useful to work from, as they relate directly to your specification.

To help with drawings

Freeman and Bracegirdle (1966) *An Atlas of Histology*, Heinemann Educational

Freeman and Bracegirdle (1976) *An Advanced Atlas of Histology*, Heinemann

Bracegirdle and Miles (1971) *Atlas of plant structure Vol I*, Heinemann Educational

Bracegirdle and Miles (1973) *Atlas of plant structure Vol II*, Heinemann Educational

Shaw, Lazell and Foster (1965) *Photomicrographs of the flowering plant*, Longman

Shaw, Lazell and Foster (1968) *Photomicrographs of the non-flowering plant*, Prentice Hall Press

Websites

Make a habit of checking www.wjec.co.uk or www.eduqas.co.uk for specific information about the course you are taking and for digital resources to support your learning. Use other websites with care. Many standard experiments are performed and discussed online. There are of course dedicated revision websites as well as websites devoted to single aspects of the course. Use these wisely and remember that just because you have seen it online, it doesn't make it true or correct and it doesn't necessarily mean it has been explained to a suitable depth for A Level students.

Use a search engine to find the organ or organism you would rather not dissect or the experiment you need help understanding. Or try these:

Simulations: many people do not want to dissect material of animal origin, indeed it is not even required at some medical schools. Instead students use web-based simulations. Some even have sound effects.
Frog dissection. tinyurl.com/aa4f7k and www.froguts.com

Videos: once students can overcome the yuck factor, watching videos can be very useful because they are made to enhance the important aspects. YouTube is full of eyes, hearts, kidneys and so on. The theory has so much more meaning when you have seen the real thing.
Heart dissection instructions and video. tinyurl.com/blvkmjd
Cow eye dissection video. tinyurl.com/cg7gpdg
Kidney dissection. tinyurl.com/d4htx73

Genetics simulation: try this exercise on the genetics of dragons to enhance your understanding of genotype and phenotype, dominant and recessive alleles and sex linkage. tinyurl.com/j22pbos

 Pointer

If you missed an experiment or you would like to see one again, search YouTube. It is bound to be there.

Journals

Biological Sciences Review is published by Philip Allan four times a year. It is aimed specifically at A Level students and has well-written and well-illustrated articles about topics in Biology and also has articles referring to examination technique.

NewScientist has excellent articles and even if you find it all too much, at least read the table of contents so that you know what is going on in the world of science, and look at the cartoons.

C.2 Use of language, spelling, punctuation and grammar

When you write under exam conditions it is most important that what you write is what you mean and that it is understood by your examiner to be what you mean. Scientific terminology is useful because the technical words have specific meanings which are unambiguous. In addition to the specialist vocabulary, your normal use of English must also convey exactly what you mean.

Spelling

>> *Pointer*

Language changes over time, but at any one time, a spelling is generally either right or wrong. It's better to be right.

Spelling must be accurate because there are many similar words with quite different meanings. Thymine and thiamine are two very different molecules. Meiosis and mitosis are quite different processes. Kew Gardens is not spelled Cue. In field work, you use a quadrat, not a quadrant. Your examiner will give you the benefit of the doubt in many cases, but where there are such similar words to distinguish, the spelling must be accurate.

Punctuation

Punctuation is important. Sometimes people write a whole page with no full stops and no capital letters. It is not up to the examiner to work out where your sentences begin and end. Keep your sentences short, avoid sub-clauses and you will not lose track of your thoughts.

Pronouns

>> *Pointer*

Take care when you use pronouns, especially 'it'.

A pronoun technically refers to the preceding noun. Sometimes people think faster than they write and use the word 'it' without realising the noun it refers to is so far back that the sentence becomes meaningless. Unless you are most confident, avoid the word 'it'. Your examiner will not try to guess what 'it' means. Your meaning must be clear.

Apostrophes

>> *Pointer*

Remember it's = it is; its is possessive.

Many people do not realise that an apostrophe can mean that a letter has been left out, e.g. 'don't' means 'do not'. A common error in this connection is the word 'its'. With an apostrophe, it's, it means it is. The i has been left out. Without the apostrophe 'its' is possessive, e.g. Look at the sky. Its colour is blue.

An apostrophe is not used when you make a plural, e.g. one dog, two dogs.

Clichés

Avoid clichés. Depending on your experiment, you may tend to slip into common phrases that are often a substitute for thought, such as 'playing God', 'slippery slope' and 'designer babies', although these particular examples are unlikely to crop up in your write up.

Miscellaneous misuse

- The word 'amount' is a problem. It has many meanings including mass, concentration, number and volume. So if you find yourself writing the word 'amount' cross it out immediately and write the word you really mean. The only exception to this rule is when you are discussing trampling.
- Another word to avoid is 'things'. It's far too vague and suggests you haven't bothered to think through carefully enough what you actually mean.
- Remember that you put your solutions in test tubes, not in testubes.
- You will write the word dependent and independent many times throughout your A Level career. Make sure the third letter from the end is an e.
- If you have a controlled variable, a factor is kept constant and so you control it. But a control experiment shows that it is the independent variable that is causing the change in the dependent variable. Do not confuse control and controlled.
- When you are using very hot water you have a hazard because it could burn your skin. But it will scald you, not scold you. Many people seem to think the water will tell them off.
- If you want to assess the colour of a liquid, use a colorimeter, not a calorimeter.
- This is Britain. If you use Americanisms but your meaning is clear, you will not be penalised. However, you are trying to write a well-produced piece of English prose so use English as spoken in this country. 'Got' not 'gotten', 'colour' not 'color', 'transport' not 'transportation', 999 not 911.

Handwriting

Your examiner will have hundreds of scripts to mark. The examiner is on your side and wants to give you marks but if your writing is illegible, however brilliant your work, you cannot get marks for it. Please take care. For example, make sure you distinguish a from o, otherwise lactase looks like lactose and it is not up to your examiner to guess which you mean. Cross each t and dot each i. Put spaces between words. Look at a Year 1 writing book and practise the exercises you haven't thought about for twelve years. You know it makes sense.

Little messages

Some candidates write a little note to the examiner at the end. It is not necessary to write, 'Thank you for marking my script', although we appreciate the gesture. It is not appropriate, however, to tell us your life story, girlfriend problems or information about your hangover. Stick to the Biology.

 Pointer

A beautifully written script with perfect spelling and punctuation and every paragraph separate and distinct is all that is needed.

D: Assessment of practical work

Our knowledge of biology is based on observations and so practical activities have always been very important when studying living organisms. There are many different aspects to practical work and by developing your practical skills throughout the course, you will meet the criteria set by your examination board.

D.1 Practical techniques

》 Pointer

You could be asked about any of these techniques in a written examination.

There are 12 practical techniques that students at AS and A Level Biology are expected to master. The practical tasks that you do throughout your course will provide you with many opportunities to practise them:

A. Use appropriate apparatus to record a range of measurements, including mass, time, volume, temperature, length and pH, e.g. when measuring water potential by change in mass of plant tissue.

B. Use appropriate instrumentation, e.g. a colorimeter when assessing the stability of beetroot membranes at different temperatures; a potometer when assessing the rate of water uptake into a transpiring plant shoot.

C. Use laboratory glassware apparatus for a variety of lab techniques, e.g. in making serial dilutions of a bacterial suspension; in investigating the effect of pH or temperature on enzyme activity.

D. Use a light microscope at high and low power, including the use of an eyepiece graticule, e.g. when measuring the diameter of a pollen grain.

E. Produce annotated scientific drawings, e.g. when examining the structure of a mammalian ovary; when examining a T.S. anther to observe the stages of meiosis.

》 Pointer

In written examination, you might be asked to do calculations relating to microscopic examination.

F. Use qualitative reagents to identify biological molecules, e.g. using Benedict's reagent to ascertain the presence of reducing sugars in a solution.

G. Separate biological compounds, e.g. using thin layer or paper chromatography to separate chloroplast pigments; using electrophoresis to separate proteins in a seed extract.

H. Safely and ethically use organisms to measure plant or animal responses and physiological functions, e.g. when investigating biodiversity in a habitat; when using locusts to assess gas exchange in a respirometer.

I. Use microbiological aseptic technique, including the use of agar plates and broth, e.g. when investigating the effect of antibiotics on *Micrococcus*; when assessing bacterial population growth in milk as it ages.

J. Safely use instruments for dissection of an animal or a plant organ, e.g. in dissecting a fish head to understand fish gas exchange; when dissecting flowers to compare strategies for wind and insect pollination.

K. Use sampling techniques in field work, e.g. when investigating continuous variation in a population; when investigating biodiversity in a habitat.

L. Use ICT in computer modelling, data logging or processing data, e.g. investigating factors affecting photosynthesis; modelling the effects of predator and prey numbers on their respective population numbers.

D.2 Specified practical work

The table below shows the experiments that are cited in your specification, and you are expected to be familiar with them. Understanding how to do these experiments and how to analyse and evaluate their results will enhance your biological understanding. You may be asked about these experiments in a written examination and, if you are to be assessed by WJEC, you may be required to perform one of these methods, or one very similar, in your practical examination.

There are also practical activities cited within the text of your specification, rather than listed separately at the end of each section, e.g. the histological examination of ovary and testis in the section 'Sexual Reproduction in Humans'. These may also provide the basis for examination questions.

	Specification section	Specified practical work
AS / Year 1	Chemical elements are joined together to form biological compounds	Food tests to include: iodine-potassium iodide test for starch; Benedict's test for reducing and non-reducing sugars; biuret test for protein; emulsion test for fats and oils
	Cell structure and organisation	Calibration of the light microscope at low and high power, including calculation of actual size of a structure and the magnification of a structure in a drawing
		Preparation and scientific drawing of a slide of living cells, e.g. onion/rhubarb/*Amoeba* including calculation of actual size and magnification of drawing
	Cell membranes and transport	Determination of water potential by measuring changes in mass or length
		Determination of solute potential by measuring the degree of incipient plasmolysis
		Investigation into the permeability of cell membranes using beetroot
	Biological reactions are regulated by enzymes	Investigation into the effect of temperature or pH on enzyme activity
		Investigation into the effect of enzyme or substrate concentration on enzyme activity
	Nucleic acids and their functions	Simple extraction of DNA from living material

Specification section		Specified practical work
	Genetic information is copied and passed on to daughter cells	Scientific drawing of cells from slides of root tip to show stages of mitosis
		Scientific drawing of cells from prepared slides of developing anthers to show stages of meiosis
	All organisms are related through their evolutionary history	Investigation into biodiversity in a habitat
A2 / Year 2	Adaptations for gas exchange	Investigation into stomatal numbers in leaves
		Dissection of fish head to show the gas exchange system
		Scientific drawing of a low power plan of a prepared slide of T.S. dicotyledon leaf, e.g. *Ligustrum* (privet), including calculation of actual size and magnification of drawing
	Adaptations for transport	Investigation into transpiration using a simple potometer
		Scientific drawing of a low power plan of a prepared slide of T.S artery and vein, including calculation of actual size and magnification of drawing
		Dissection of mammalian heart
	Photosynthesis uses light energy to synthesise organic molecules	Investigation into the separation of chloroplast pigments by chromatography
		Investigation into factors affecting the rate of photosynthesis
		Investigation into the role of nitrogen and magnesium in plant growth
	Respiration releases chemical energy in biological processes	Investigation into factors affecting the rate of respiration in yeast
	Population size and ecosystems	Investigation into the abundance and distribution of organisms in a habitat
	Homeostasis and the kidney	Dissection of kidney
	Sexual reproduction in plants	Investigation of the digestion of starch agar using germinating seeds
		Dissection of wind- and insect-pollinated plants
		Scientific drawing of a low power plan of a prepared slide of an anther, including calculation of actual size and magnification of drawing
	Inheritance	Experiment to illustrate gene segregation including the use of the chi-squared test in assessing the significance of genetic outcomes
	Variation and evolution	Investigation of continuous variation in a species (including use of Student's t-test)

• • • **Practical tip** • • •

Watch these experiments again on YouTube as part of your exam preparation.

D.3 Your practical assessment

Assessment of practical work accounts for 15% of your final AS or A Level grade. There are two models for practical assessment, depending on which examination board is assessing you:

- WJEC: In written examinations and in a practical examination.
- Eduqas: In written examinations and by an endorsement of your practical work.

Written examinations (WJEC and Eduqas)

At both AS and A Level, you will be asked questions about practical work in the written examinations, so it is important that you take every opportunity to do it. There are many types of question that can test your experience in the lab, and your understanding of experimental design and analysis. You may be asked about familiar experiments, such as those cited in the specification. There may also be questions about novel practical situations. The skills you have developed during the course will allow you to answer these.

Some examples are given here:

Question topic	Example
Magnification and measurement	In calculating the magnification of a photograph when you have been given an actual length.
Design an experiment	Given a null hypothesis, e.g. there is no significant difference between the maximum diameter of ground ivy leaves grown at 12°C and 22°C, you should be able to design an experiment to test. You should give: • the independent variable • the dependent variable • at least two controlled variables • an explanation of how you would make your results reliable • an explanation of how you would make your results accurate.
The purpose of a step in a protocol	In separating chloroplast pigments by chromatography, you should be able to explain why you use a hairdrier to dry the pigment spot as soon as it has been applied to the chromatography paper or plate.
Describe a procedure	You will only be asked to describe a particular procedure, such as streak plating, making a root tip squash or serial dilution, that has been identified in your specifications.
Modify a procedure	In an experiment to determine a rate of reaction, modifying a procedure might involve dissolving reactants in pH7 buffer, to maintain the pH, rather than water, or running the reactions in a water bath, to maintain constant temperature, instead of on the lab bench.
Ethical treatment of animals	If you are estimating a population size with the Lincoln index, any animals that you catch should be marked as unobtrusively as possible, with a non-toxic mark; they should be returned as close as possible to where you sampled them, as soon as possible.
Valid results	An experiment is valid if it measures what it sets out to measure. For an experiment to be valid, the readings must be reliable and accurate.

Question topic	Example
Reliability	To increase the reliability of a mean, more readings should be taken. If there is clearly an anomalous reading, it is justifiable to remove it before calculating the mean, as long as you explain what you have done and why.
Sources of error	Judging a colour change, e.g. when methylene blue has decolorised, is subjective. Using a control tube or colour chart as a reference would reduce the error in the timing. If your stopwatch measures to 0.1 seconds, using one correct to 0.01 seconds is not an improvement because the colour change is gradual.
Improvements	In a field work experiment assessing the mean diameter of a species of limpet at different heights up the shore, an improvement would be to use a field guide to ensure that the same species is always being sampled.
Increase precision	The accuracy of any reading is limited by the calibration on the measuring equipment. If you are using a respirometer and collecting carbon dioxide in a syringe barrel calibrated in cm^3, a reading of volume is only accurate to the nearest 0.5 cm^3. Using a syringe calibrated to 0.1 cm^3 increases the accuracy.

>> *Pointer*

Some people are nervous in practical exams. You don't need to be. The experiment is designed to use techniques with which you are familiar and you are given all the equipment you need.

Practical examination (WJEC)

If your A Level examining board is WJEC, towards the end of your A Level course, you will have a practical examination. It has two parts:

- A practical task lasting two hours. You will be given written instructions for carrying out an experiment. You will be provided with all the equipment you need and an outline method, but you will need to choose which variables you control and how many replicate readings to take. Your teacher will confirm that you have set up and carried out the procedure correctly during the test. You will construct a suitable table of readings and calculations and use them to plot a graph. You will also answer questions about the experiment.
- A practical analysis task lasting one hour. You will be given information about a biological phenomenon and data to illustrate it. You will be asked questions about it, in a practical context. You will also be asked to perform calculations to do a statistical test on the data and you will answer questions that test your ability to analyse the evidence. Formulae and tables of critical values will be given to you, but you must know how to substitute into the formulae and how to interpret the test statistic that you have calculated.

Practical endorsement (Eduqas)

At A Level, an endorsement of your practical will form a part of your assessment. Your teacher will observe you doing practical work throughout the course and will note how you recorded, processed and interpreted your results. If you have consistently and routinely achieved the criteria explained on 81–83, you will receive the practical endorsement for your work.

Throughout the course, you will keep a lab book, which is the written evidence of your work. The lab book is a working document. It will include:

- a record of what you have done in practical sessions
- the dates on which you achieve the competencies, and the experiments in which you achieved them.

The lab book is, therefore, a record of the development of your practical skills. Your teacher will not necessarily mark it, but may annotate your work with advice, e.g. on table construction or drawing technique. By incorporating the ideas into future pieces of work, your lab book shows that you have acted on that advice.

On your A Level certificate, next to your grade will appear the words 'pass' or 'not classified', depending on whether or not your practical work received the endorsement. Your final A Level grade is derived entirely from your written examination papers, and whether or not you received the practical endorsement has no impact on it.

>> *Pointer*
In your lab book, construct a table of results before you start taking readings and write the readings directly into the table as your experiment progresses.

Criteria for Eduqas practical endorsement

The progress you make in practical work will be assessed throughout your course in accordance with the five 'Common Practical Assessment Criteria' (CPAC). The experiments in which these criteria are assessed will be done when you are at the relevant stage of the course. You will be assessed for each criterion several times so that your teacher can see that you routinely and consistently achieve each one. Your teacher will tell you which criterion or criteria are being assessed at the start of a practical session. In most cases, only one or two criteria will be assessed for any particular task.

>> *Pointer*
Make sure you understand what you must do to achieve each of the five CPAC statements.

For your A2 practical assessment, to achieve the five CPACs, you have to demonstrate your problem-solving and investigational skills. To do this, you will need to demonstrate certain skill areas encompassed by the criteria:

- Planning and decision making
- Manipulation, measurement and observation
- Presentation of data and observations
- Analysis of data and conclusions
- Evaluation of procedures and data

You are not likely to write about every one of these in a full lab report for each experiment that you do; your teacher will often ask you to write up only some. You will, however, have many opportunities to show that you are developing these skills and that you become consistently competent at meeting the five criteria. It is, therefore, important that you understand the criteria that are being assessed. They are described below:

CPAC 1: Follows written procedures

When you have an experimental task to perform, you will be given some indication of how to do it, such as verbal instructions, a written sheet of prose, a flow diagram or a list of points to consider. The amount of detail you are given will depend on the task and on your previous experience. At the start of the course, your teacher is likely to talk through the method and variables in detail. By the end of the course, you may be given only bullet points outlining the method, and you may be expected to choose the variables. You would, however, be alerted to any particular health and safety issue. A piece of equipment that you had never seen before, such as a colorimeter, or a technique you had never previously used, such as chromatography, would, of course, be demonstrated.

>> *Pointer*
Read instructions carefully so you are clear what your independent and dependent variables are.

But when equipment is familiar, such as test tubes and syringes for making a dilution, you should be able to follow instructions directly.

Your teacher will assess how well you follow the instructions, in whatever form they take. Do not, however, be afraid to ask questions if you are unsure. You will be assessed on how you progress through the course, and asking questions is one way to make progress.

CPAC 2: Applies investigative approaches and methods when using instruments and equipment

As your understanding of Biology and your experience in lab work increase, you will be given fewer specific instructions for practical tasks, which will allow you to develop confidence in solving problems without being prompted. Your teacher will assess the extent to which you increasingly generate your own ideas about how to design experiments. If, for example, you were asked to investigate the plants in a habitat, you might choose one of several methods, e.g.:

- measuring the density of the dominant plant
- calculating a diversity index by assessing all the plant species present
- measuring the percentage area cover of the ground flora.

These are all valid investigations and you should be able to support your choice of experiment and explain why you made it.

Your assessment of CPAC 2 may be for part of an experiment only, rather than a full investigation.

>> **Pointer**

Make sure you can identify the most significant hazard in an experiment and take steps to reduce the risk it poses, e.g. the potential for allergies to yeast is more important than the risk of breaking glassware.

CPAC 3: Safely uses a range of practical equipment and materials

The way you plan and carry out your experiments must show that you can identify hazards and can take steps to avoid risk. You should be able to recognise and act on symbolic hazard warnings on containers of chemicals and on equipment. Your lab work might put you in the situation of dealing with unexpected hazards, for example, a piece of glassware breaking; you would have to show that you can act safely in such an event. The way you arrange apparatus is important so, for example, you would take care to avoid reaching across a flame to pick up a Petri dish and you would not place apparatus near the edge of the lab bench. You would take suitable measures pouring potentially harmful liquids, such as hydrogen peroxide, and when using sharp instruments, such as a scalpel. This way you protect yourself and other people.

If you are doing a field work experiment, you must show that you are aware of safety, of laws protecting wildlife and of specialist advice published by conservation groups. You would not treat any animal in any way that might cause it pain or distress and you would not remove plants that have legal protection.

You are not required to write a full risk assessment, but it is good practice to consider potential hazards and their associated risks, to avoid harm.

CPAC 4: Makes and records observations

The way you record your readings must show that you have been methodical in your observations. Tables should be constructed with suitable column headings and units. Standard practice places the independent variable in the first column, with replicate readings of the dependent variable in the following columns. All raw data must be shown, with readings given to an appropriate number of decimal places in relation to the calibrations on the equipment you have used. You would show that you understand how many replicates are needed. You must explain clearly your reasoning if you discard a reading that you consider to be anomalous.

If the record is a drawing, rather than numerical data, you should show the correct proportions and a scale. Individual cells are shown in high power drawings only, not in low power plans.

CPAC 5: Researches, references and reports

Your teacher will assess your ability to undertake and report research associated with an experiment. Your written report would show that you can manipulate the raw data, either with a spreadsheet or by using a calculator, although using a data logger would not be appropriate in this context. You should evaluate your results and draw a sensible conclusion, based on your knowledge of Biology. If the experiment has been to calculate a parameter, you might compare your calculated value with a theoretical value that you have found in your research. Conclusions should be supported with references to journals or websites, which are referenced in such a way that they can be retrieved, with the date they were accessed.

Pointer

If you want to change a number in a table, cross it out and rewrite it. Do not overwrite it, because then it is hard to read what you have written.

Pointer

Take care to cite every reference you use. Avoid any hint of plagiarism.

Glossary

Sometimes an exam question will ask about practical work. You may get a question that starts 'What is the meaning of the term ...?' This glossary gives concise meanings of many of the technical terms used in practical work.

Section	Term	Meaning
Microscopy	Magnification	Ratio of image size : object size
	Resolution	Smallest distance between two points that can be distinguished
	Calibration	Calculation of actual length of scale on eyepiece graticule at a given objective lens magnification
	Graticule	Scale mounted inside eyepiece
	Stage micrometer	Slide accurately ruled showing sub-divisions of a 1 cm line
	smu	Stage micrometer unit
	epu	Eyepiece unit; graduation on eyepiece graticule
	Photomicrograph	Photograph taken down a microscope
	Electron micrograph	Photograph taken using an electron microscope
Aseptic culture	Aseptic (sterile)	The absence of contaminating organisms, e.g. bacteria, fungi
How to design a lab experiment	Variable	A factor that may change
	Independent variable	A factor chosen to change; values decided by experimenter
	Dependent variable	Measured, counted or calculated factor, the value of which depends on the independent variable
	Controlled variable	A factor that is kept constant throughout the experiment, so that only the independent variable is allowed to change prior to measuring the value of the dependent variable
	Control experiment	Experiment with major factor inactivated to demonstrate its significance in generating change in the dependent variable
	Accurate	True
	Repeatable	Reliable; consistent; readings are close to one another
	Reproducible	Experiment can be set up another way and get the same results
	Q_{10}	Rate of reaction at $(t+10)°C$ / rate of reaction at $t°C$
	R_f	Distance moved by pigment / distance moved by solvent front
How to design a field work experiment	Sampling	Selection of individuals from a population to estimate the characteristics of the whole population
	Quadrat	A frame, often $0.25 m^2$ square, open or gridded, for sampling organisms
	Transect	A line or belt along which organisms are identified, to contribute to defining the characteristics of a habitat
	Density	Number of organisms per m^2 on land or per m^3 in water
	Percentage area cover	Percentage area occupied by a given species

Index